MF/1 (rev 4)

Model form of

General Conditions of Contract

including

Forms of Tender, Agreement, Sub-contract, Performance Bond
and
Defects Liability Demand Guarantee

For use in connection with

**Home or overseas contracts for the supply of electrical, electronic or mechanical plant
—with erection**

Recommended by

The Institution of Electrical Engineers

The Institution of Mechanical Engineers
and
The Association of Consulting Engineers

2000 Edition

Historical note

A "Form of Model Conditions recommended for use in connection with contracts for plant, mains and apparatus for electricity works" was originally drawn up by a committee convened by the Council of the Institution of Electrical Engineers in 1903. This was followed by:

1914
Form of Model General Conditions recommended for use in connection with contracts for electrical works.

1921, 1926, 1929, 1938
Model Form of General Conditions 'A', Home Contracts - with erection.

1925, 1928
Model Form of General Conditions 'B.2'
(Export, including Complete Erection or Supervision of Erection).

Following agreement between the Councils of the Institution of Mechanical Engineers and the Institution of Electrical Engineers, the scopes of the forms were enlarged to make them suitable for both the electrical and mechanical engineering industries, and, on this basis, the model forms were issued jointly by the two Councils.

1948
Model Form of General Conditions of Contract 'A', Home Contracts - with erection.

1954
Model Form of General Conditions of Contract 'B3' - Export Contracts (including Delivery to and Erection on Site).

As a result of an approach by The Association of Consulting Engineers, it was agreed in 1951 that the Association should adopt and join in recommending this model form, simultaneously discontinuing the issue of the "Conditions of Contract" which the Association had hitherto prepared and recommended for use by its members. Accordingly, the name of the Association now appears on the title page of the model form.

1966, 1971, 1972, 1973, 1976, 1978, 1982
Model Form of General Conditions 'A', Home Contracts - with erection.
These revisions covered: developments in practice (1966); amendments to cover S.E.T. and metrication (1971); V.A.T. (1972); V.A.T. and abolition of S.E.T. (1973); general revision (1976); amendments in response to the Unfair Contract Terms Act 1977 (1978); the Arbitration Act 1979, and abolition of the Bank of England minimum lending rate (1982).

1971, 1980, 1982
Model Form of General Conditions of Contract 'B3', Export Contracts - with erection.

1988
Model Form of General Conditions of Contract MF/1, Home or Overseas Contracts - with erection.
This was an entirely new model form, suitable for both home and export contracts, and replaced both Model Forms 'A' and 'B3'. It recognised important changes in practice, particularly in the role of the Engineer, and incorporated special sections covering sub-contracts, and electronics hardware and software. An explanatory Commentary was produced to accompany the form and explain the underlying principles.

1989, 1992
MF/1 Reprints incorporating editorial amendments.

1993
Separate Supplement, S1-MF/1, published —covering suggested Special Conditions for contracts involving measurement.

1995
MF/1 (rev 3), Home or Overseas Contracts for the Supply of Electrical, Electronic or Mechanical Plant - with Erection, published —incorporating both editorial and other amendments and a revision covering unregistered design rights and Applicable Law. Additionally, the 1993 Supplement (S1-MF/1) has been integrated into the form and further suggested Special Conditions to allow for sectional completions have been provided.

2000
2000 Edition of MF/1, MF/1 (rev 4), published —incorporating both editorial and other amendments. Additionally, the 1998 and 1999 Supplements (S2-MF/1 & S3-MF/1) and the 2000 Amendment Slip (MF/1, A/S1) to MF/1 have been integrated into the form.

ISBN 0 85296 759 4

12/00

Table of contents

		Page
Historical note		ii
Contents		iii
General conditions		

Clause		
	Definitions and interpretations	
1.1	Definitions	1
1.2	Interpretation	3
1.3	Singular and plural	3
1.4	Notices and consents	3
1.5	Headings and marginal notes	3
	Engineer and engineer's representative	
2.1	Engineer's duties	3
2.2	Engineer's representative	3
2.3	Engineer's power to delegate	3
2.4	Engineer's decisions, instructions and orders	3
2.5	Confirmation in writing	3
2.6	Disputing engineer's decisions, instructions and orders	4
2.7	Engineer to act fairly	4
2.8	Replacement of engineer	4
	Assignment and sub-contracting	
3.1	Assignment	4
3.2	Sub-contracting	4
	Precedence of documents	
4.1	Precedence of documents	4
	Basis of tender and contract price	
5.1	Contractor to inform himself fully	5
5.2	Site data	5
5.3	Site data	5
5.4	Provisional sums	5
5.5	Prime cost items	5
5.6	Prime cost items	5
5.7	Unexpected site conditions	5
	Changes in costs	
6.1	Statutory and other regulations	5
6.2	Labour, materials and transport	5
	Agreement	
7.1	Agreement	6
	Performance bond or guarantee	
8.1	Provision of bond or guarantee	6
8.2	Failure to provide bond or guarantee	6

Clause		Page
	Details confidential	
9.1	Details confidential	6
	Notices	
10.1	Notices to purchaser and engineer	6
10.2	Notices to contractor	6
10.3	Service of notices	7
	Purchaser's general obligations	
11.1	Access to site	7
11.2	Wayleaves, consents, etc.	7
11.3	Import permits, licences and duties	7
11.4	Foundations, etc.	7
11.5	Purchaser's lifting equipment	7
11.6	Utilities and power	7
11.7	Power, etc. for tests on site	7
11.8	Breach of purchaser's general obligations	7
12.1	Assistance with laws and regulations	7
	Contractor's obligations	
13.1	Contractor's general obligations	8
13.2	Manner of execution	8
13.3	Contractor's design	8
14.1	Programme	8
14.2	Form of programme	9
14.3	Approval of programme	9
14.4	Alterations to programme	9
14.5	Revision of programme	9
14.6	Rate of progress	9
15.1	Drawings	9
15.2	Consequences of disapproval of drawings	9
15.3	Approved drawings	9
15.4	Inspection of drawings	9
15.5	Foundation, etc. drawings	10
15.6	Operating and maintenance instructions	10
15.7	Purchaser's use of drawings, etc. supplied by contractor	10
15.8	Contractor's use of drawings, etc. supplied by purchaser or engineer	10
15.9	Manufacturing drawings, etc.	10
16.1	Errors in drawings, etc. supplied by contractor	10
16.2	Errors in drawings, etc. supplied by purchaser or engineer	11
17.1	Contractor's representatives and workmen	11
17.2	Objection to representatives	11
17.3	Returns of labour	11
18.1	Fencing, guarding, lighting and watching	11
18.2	Site services	11
18.3	Clearance of site	11
18.4	Opportunities for other contractors	11
19.1	Hours of work	11
19.2	No night or rest day working	12

Clause		Page
20.1	Safety	12
21.1	Extraordinary traffic	12
21.2	Special loads	12
21.3	Extraordinary traffic claims	12
21.4	Waterborne traffic	12
22.1	Setting out	12
	Inspection and testing of plant before delivery	
23.1	Engineer's entitlement to test	13
23.2	Date for test or inspection	13
23.3	Services for test or inspection	13
23.4	Certificate of test or inspection	13
23.5	Failure on test or inspection	13
24.1	Delivery	13
	Suspension of work, delivery or erection	
25.1	Instructions to suspend	14
25.2	Additional cost caused by suspension	14
25.3	Payment for plant affected by suspension	14
25.4	Disallowance of additional cost or payment	14
25.5	Resumption of work, delivery or erection	14
25.6	Effect of suspension on defects liability	15
	Defects before taking-over	
26.1	Defects before taking-over	15
	Variations	
27.1	Meaning of variation	15
27.2	Engineer's power to vary	16
27.3	Valuation of variations	16
27.4	Contractor's records of costs	16
27.5	Notice and confirmation of variations	16
27.6	Progress with variations	16
	Tests on completion	
28.1	Notice of tests	17
28.2	Time for tests	17
28.3	Delayed tests	17
28.4	Repeat tests	17
28.5	Consequences of failure to pass tests on completion	17
	Taking-over	
29.1	Taking-over by sections	17
29.2	Taking-over certificate	17
29.3	Effect of taking-over certificate	17
29.4	Outstanding work	17
30.1	Use before taking-over	18
31.1	Interference with tests	18
31.2	Tests during defects liability period	18

Clause			Page
	Time for completion		
32.1	Time for completion		18
33.1	Extension of time for completion		18
33.2	Delays by sub–contractors		18
33.3	Mitigation of consequences of delay		19
	Delay		
34.1	Delay in completion		19
34.2	Prolonged delay		19
	Performance tests		
35.1	Time for performance tests		19
35.2	Procedures for performance tests		19
35.3	Cessation of performance tests		19
35.4	Adjustments and modifications		20
35.5	Postponement of adjustments and modifications		20
35.6	Time for completion of performance tests		20
35.7	Evaluation of results of performance tests		20
35.8	Consequences of failure to pass performance tests		20
	Defects liability		
36.1	Defects after taking-over		21
36.2	Making good defects		21
36.3	Notice of defects		21
36.4	Extension of defects liability		21
36.5	Delay in remedying defects		21
36.6	Removal of defective work		21
36.7	Further tests		21
36.8	Contractor to search		22
36.9	Limitation of liability for defects		22
36.10	Latent defects		22
	Vesting of plant, and contractor's equipment		
37.1	Ownership of plant		22
37.2	Marking of plant		22
38.1	Contractor's equipment		23
38.2	Contractor's equipment on site		23
38.3	Loss or damage to contractor's equipment		23
38.4	Maintenance of contractor's equipment		23
	Certificates and payment		
39.1	Application for payment		23
39.2	Form of application		23
39.3	Issue of payment certificate		24
39.4	Value included in certificates of payment		24
39.5	Adjustments to certificates		24
39.6	Corrections to certificates		24
39.7	Withholding certificate of payment		24
39.8	Effect of certificates of payment		24
39.9	Application for final certificate of payment		24
39.10	Value of final certificate of payment		25
39.11	Issue of final certificate of payment		25

Clause			Page
39.12	Effect of final certificate of payment		25
39.13	No effect in the case of gross misconduct		25
40.1	Payment		25
40.2	Delayed payment		25
40.3	Remedies on failure to certify or make payment		26
	Claims		
41.1	Notification of claims		26
41.2	Allowance for profit on claims		26
41.3	Purchaser's liability to pay claims		26
	Patent rights, etc.		
42.1	Indemnity against infringement		27
42.2	Conduct of proceedings		27
42.3	Purchaser's indemnity against infringement		27
42.4	Infringement preventing performance		27
	Accidents and damage		
43.1	Care of the works		28
43.2	Making good loss or damage to the works		28
43.3	Damage to works caused by purchaser's risks		28
43.4	Injury to persons or damage to property whilst contractor has responsibility for care of the works		28
43.5	Injury to persons or damage to property after responsibility for care of the works passes to purchaser		28
43.6	Accidents or injury to workmen		28
43.7	Claims in respect of injury to persons or damage to property		29
	Limitations of liability		
44.1	Mitigation of loss		29
44.2	Indirect or consequential damage		29
44.3	Limitation of contractor's liability		29
44.4	Exclusive remedies		29
	Purchaser's risks		
45.1	Purchaser's risks		29
	Force majeure		
46.1	Force majeure		30
46.2	Notice of force majeure		30
46.3	Termination for force majeure		30
46.4	Payment on termination for force majeure		30
	Insurance		
47.1	Insurance of works		31
47.2	Extension of works insurance		31
47.3	Application of insurance monies		31
47.4	Third party insurance		31
47.5	Insurance against accident, etc. to workmen		32
47.6	General insurance requirements		32
47.7	Exclusions from insurance cover		32
48.1	Remedy on failure to insure		32
48.2	Joint insurances		32

Clause		Page
	Contractor's default	
49.1	Contractor's default	32
49.2	Valuation at date of termination	33
49.3	Payment after termination	33
50.1	Bankruptcy and insolvency	33
	Purchaser's default	
51.1	Notice of termination due to purchaser's default	34
51.2	Removal of contractor's equipment	34
51.3	Payment on termination due to purchaser's default	34
	Disputes and arbitration	
52.1	Notice of arbitration	34
52.2	Performance to continue during arbitration	35
52.3	Arbitrator's powers	35
52.4	Joinder	35
52.5	Arbitration rules	35
	Sub–contractors, etc.	
53.1	Sub–contractors, servants and agents	35
	Applicable law	
54.1	Applicable law	35

Appendix to general conditions

	Page
Appendix	36

Special conditions

	Page
Aide-mémoire to their preparation	37
Additional special conditions for use in contracts involving the incidental supply of hardware and software	42
Additional special conditions for use in contracts where certification for payment and payments are to be determined in whole or in part by measurement	47
Additional special conditions for use where the contract is to provide sectional completion and damages for delay in completion of sections	51
Additional special conditions for use in contracts which are subject to the Housing Grants, Construction and Regeneration Act 1996	52
Additional special conditions for use in contracts which are subject to the Contracts (Rights of Third Parties) Act 1999	55

	Page
Form of Sub-Contract	56
Form of Tender	64
Form of Agreement	66
Form of Performance Bond	68
Form of Defects Liability Demand Guarantee	70
Form of Notice of Delegation of Authority	72

	Page
Variation Order	73
Taking-Over Certificate	74

Index to general conditions
Index   75

GENERAL CONDITIONS

Definitions and interpretations

Definitions **1.1** In construing the Contract the following words and expressions shall have the following meanings hereby assigned to them.

1.1.a 'Purchaser' means the person named as such in the Special Conditions and the legal successors in title to the Purchaser but not (except with the consent of the Contractor) any assignee of the Purchaser.

1.1.b 'Contractor' means the tenderer whose Tender has been accepted by the Purchaser and the legal successors in title to the Contractor but not any assignee of the Contractor.

1.1.c 'Sub–Contractor' means any person (other than the Contractor) named in the Contract for the supply of any part of the Works or any person to whom any part of the Contract has been sub–let with the consent of the Engineer, and the Sub–Contractor's legal successors in title, but not any assignee of the Sub–Contractor.

1.1.d 'Engineer' means the person appointed by the Purchaser to act as Engineer for the purposes of the Contract and designated as such in the Special Conditions or, in default of any appointment, the Purchaser.

1.1.e 'Engineer's Representative' means any assistant of the Engineer appointed from time to time to perform the duties delegated to him under clause 2 (Engineer and engineer's representative) hereof.

1.1.f 'Conditions' means these general conditions and the Special Conditions.

1.1.g 'Contract' means the agreement between the Purchaser and the Contractor (howsoever made) for the execution of the Works including the Letter of Acceptance, the Conditions, the Specification and the drawings (if any) annexed thereto and such schedules as are referred to therein and the Tender.

1.1.h 'Contract Price' means the sum stated in the Contract as the price payable to the Contractor for the execution of the Work.

1.1.i 'Contract Value' means such part of the Contract Price, adjusted to give effect to such additions or deductions as are provided for in the Contract, other than under sub–clause 6.2 (Labour, materials and transport), as is properly apportionable to the Plant or work in question. In determining Contract Value the state, condition and topographical location of the Plant, the amount of work done and all other relevant circumstances shall be taken into account.

1.1.j 'Cost' means all expenses and costs incurred including overhead and financing charges properly allocable thereto with no allowance for profit.

1.1.k 'Tender' means the Contractor's priced offer to the Purchaser for the execution of the Works.

1.1.l 'Letter of Acceptance' means the formal acceptance by the Purchaser of the Tender incorporating any amendments or variations to the Tender agreed by the Purchaser and Contractor.

1.1.m 'Time for Completion' means the period of time for completion of the Works or any Section thereof as stated in the Contract or as extended under sub–clause 33.1 (Extension of time for completion) calculated from whichever is the later of:

Copyright © 2000 Institution of Electrical Engineers

(a) the date specified in the Contract as the date for commencement of the Works;

(b) the date of receipt of such payment in advance of the commencement of the Works as may be specified in the Contract,

(c) the date on which any necessary legal, financial or administrative requirements specified in the Contract as conditions precedent to commencement have been fulfilled.

1.1.n 'Contractor's Equipment means all appliances or things of whatsoever nature required for the purposes of the Works but does not include Plant, materials or other things intended to form or forming part of the Works.

1.1.o 'Plant' means machinery, computer hardware and software, apparatus, materials, articles and things of all kinds to be provided under the Contract other than Contractor's Equipment.

1.1.p 'Works' means all Plant to be provided and work to be done by the Contractor under the Contract.

1.1.q 'Section of the Works' or 'Section' means the parts into which the Works are divided by the Specification.

1.1.r 'Programme' means the programme referred to in clause 14 (Programme).

1.1.s 'Specification' means the specification of the Works annexed to or included in the Contract including any modifications thereof made under clause 21 (Variations).

1.1.t 'Special Conditions' means the alterations to these general conditions and such further conditions as may be specified and identified as the Special Conditions in the Contract.

1.1.u 'Site' means the actual place or places, provided or made available by the Purchaser, to which Plant is to be delivered or at which work is to be done by the Contractor, together with so much of the area surrounding the same as the Contractor shall with the consent of the Purchaser actually use in connection with the Works otherwise than merely for the purposes of access.

1.1.v 'Tests on Completion' means the tests specified in the Contract (or otherwise agreed by the Purchaser and the Contractor) which are to be made by the Contractor upon completion of erection and/or installation before the Works are taken over by the Purchaser.

1.1.w 'Performance Tests' means the tests (if any) detailed in the Specification or in a performance tests schedule otherwise agreed between the Purchaser and the Contractor, to be made after the Works have been taken over to demonstrate the performance of the Works.

1.1.x 'Defects Liability Period' has the meaning assigned by sub–clause 36.1 (Defects after taking-over).

1.1.y 'Purchaser's Risks' has the meaning assigned by clause 45 (Purchaser's risks).

1.1.z 'Force Majeure' has the meaning assigned by sub-clause 46.1 (Force majeure).

1.1.aa 'Appendix' means the Appendix to these general conditions.

1.1.bb 'writing' means any hand-written, type-written or printed statement.

1.1.cc 'day' means calendar day.

1.1.dd 'week' means any period of seven days.

1.1.ee 'month' means calendar month.

Interpretation

1.2 Words importing persons or parties shall include firms, corporations and any organisation having legal capacity.

Singular and plural

1.3 Words importing the singular only also include the plural and vice versa where the context requires.

Notices and consents

1.4 Wherever in the Conditions provision is made for the giving of notice or consent by any person, unless otherwise specified such notice or consent shall be in writing and the word 'notify' shall be construed accordingly.

Any consent required of a party or the Engineer shall not be unreasonably withheld.

Headings and marginal notes

1.5 The headings or marginal notes in the Conditions shall not be deemed part thereof or be taken into consideration in the interpretation or construction thereof or of the Contract.

Engineer and engineer's representative

Engineer's duties

2.1 The Engineer shall carry out such duties in issuing certificates, decisions, instructions and orders as are specified in the Contract.

If the Engineer is required, under the terms of his appointment by the Purchaser, to obtain the prior specific approval of the Purchaser before exercising any of his duties under the Contract, particulars of such requirements shall be set out in the Special Conditions.

Engineer's representative

2.2 The Engineer's Representative shall be responsible to the Engineer and shall watch and supervise the Works, and test and examine any Plant or workmanship employed in connection therewith.

The Engineer's Representative shall have only such further authority as may be delegated to him by the Engineer under sub-clause 2.3 (Engineer's power to delegate).

Engineer's power to delegate

2.3 The Engineer may from time to time delegate to the Engineer's Representative any of his duties and he may at any time revoke such delegation.

Any delegation or revocation shall be in writing. The Engineer shall furnish to the Contractor and to the Purchaser a copy of any such delegation or revocation. No such delegation or revocation shall have effect until a copy thereof has been delivered to the Contractor.

Any written decision, instruction, order or approval given by the Engineer's Representative to the Contractor in accordance with such delegation shall have the same effect as though it had been given by the Engineer.

If the Contractor disputes or questions any decision, instruction or order of the Engineer's Representative he may refer the matter to the Engineer who shall confirm, reverse or vary the decision, instruction or order in accordance with sub-clause 2.6 (Disputing engineer's decisions, instructions and orders).

Engineer's decisions, instructions and orders

2.4 The Contractor shall proceed with the Contract in accordance with the decisions, instructions and orders given by the Engineer in accordance with the Conditions.

Confirmation in writing

2.5 The Contractor may require the Engineer to confirm in writing any decision, instruction or order of the Engineer which is not in writing. The Contractor shall make such request without undue delay. Such a decision,

Disputing engineer's decisions, instructions and orders

instruction or order shall not be effective until written confirmation thereof has been received by the Contractor.

2.6 If the Contractor by notice to the Engineer within 21 days after receiving any decision, instruction or order of the Engineer in writing or written confirmation thereof under sub-clause 2.5 (Confirmation in writing), disputes or questions the same, giving his reasons for so doing, the Engineer shall within a further period of 21 days by notice to the Contractor and the Purchaser with reasons, confirm, reverse or vary such decision, instruction or order.

If either the Contractor or the Purchaser disagrees with such decision, instruction or order as confirmed, reversed or varied he shall be at liberty to refer the matter to arbitration within a further period of 21 days. In the absence of such a reference to arbitration within the said period of 21 days such decision, instruction or order of the Engineer shall be final and binding on the parties.

Engineer to act fairly

2.7 Wherever by the Conditions the Engineer is required to exercise his discretion:

- by giving his decision, opinion or consent
- by expressing his satisfaction or approval
- by determining value
- or otherwise by taking action which may affect the rights and obligations of either of the parties,

he shall exercise such discretion fairly within the terms of the Contract and having regard to all the circumstances.

Replacement of engineer

2.8 The Purchaser shall not appoint any person to act with the Engineer or in replacement of the Engineer without the Contractor's prior consent.

Assignment and sub-contracting

Assignment

3.1 The Contractor shall not assign the benefit of the Contract in whole or in part or any of his obligations under the Contract. A charge in favour of the Contractor's bankers of any monies due under the Contract, or the subrogation of insurers to the Contractor's rights, shall not be considered an assignment.

Sub-contracting

3.2 Except where otherwise provided by the Contract the Contractor shall not sub-contract any part of the Works without the prior consent of the Engineer.

The Contractor shall however not require such consent to place contracts for minor details nor for purchases of materials nor for any part of the Works of which the manufacturer or supplier is named in the Contract.

The Contractor shall be responsible for the acts, defaults and neglects of any Sub-Contractor, his agents, servants or workmen as fully as if they were the acts, defaults and neglects of the Contractor, his agents, servants or workmen.

Precedence of documents

Precedence of documents

4.1 Unless otherwise provided in the Contract the Conditions as amended by the Letter of Acceptance shall prevail over any other document forming part of the Contract and in the case of conflict between the general conditions and the Special Conditions the Special Conditions shall prevail. Subject thereto the Specification shall prevail over any other document forming part of the Contract.

Basis of tender and contract price

Contractor to inform himself fully

5.1 The Contractor shall be deemed to have inspected the Site and to have satisfied himself as far as can reasonably be done as to the conditions of and all circumstances affecting the Site (including any safety regulations of the Purchaser applicable thereto), if access has been made available to him, and to have examined the Conditions and Specification, with such drawings, schedules, plans and information as may be annexed thereto or referred to therein.

Site data

5.2 The Tender shall be deemed to have been based on such data on climatic, hydrological, soil and general conditions of the Site and for the operation of the Works as the Purchaser or the Engineer has made available in writing to the Contractor for the purposes of the Tender. The Contractor shall be responsible for his own interpretation of such data.

5.3 The Contractor shall be responsible for any misunderstanding or incorrect information however obtained except information provided in writing by the Purchaser or the Engineer.

Provisional sums

5.4 Provisional sums included in the Contract Price shall be expended or used as the Engineer may in writing direct and not otherwise. In so far as a provisional sum included in the Contract Price is not expended or used it shall be deducted from the Contract Price.

Prime cost items

5.5 All sums included in the Contract Price in respect of prime cost items shall be expended or used as the Engineer may in writing direct and not otherwise. To the net amount paid by the Contractor in respect of each prime cost item there shall be added the percentage thereof stated in the Appendix.

5.6 The Contractor shall have no responsibility for work done or Plant supplied by any other person in pursuance of directions given by the Engineer under the preceding sub-clause unless the Contractor shall have approved in writing the person by whom such work is to be done or such Plant is to be supplied and the Plant, if any, to be supplied.

Unexpected site conditions

5.7 In the case of work underground or involving excavation where the actual conditions of the ground are not stated in the Contract or if rock, rocky soil, solid chalk, water, running sand, slag, pipes, concrete or other obstructions are found, and such conditions or obstructions could not reasonably have been ascertained from an inspection of the Site by the Contractor before he prepared the Tender or from information made available to the Contractor for the purposes of the Tender, or if it should be necessary to leave in timber or provide support for existing work (such necessity not having been indicated in the Contract), the Contractor shall forthwith inform the Engineer of the obstructions or hazards encountered and obtain the approval of the Engineer to the steps he proposes to take to deal with the same. If the Contractor in taking such steps incurs extra Cost, such Cost shall be added to the Contract Price.

Changes in costs

Statutory and other regulations

6.1 If the Cost to the Contractor of performing his obligations under the Contract shall be increased or reduced by reason of the making after the date of the Tender of any law or of any order, regulation or bylaw having the force of law that shall affect the Contractor in the performance of his obligations under the Contract, the amount of such increase or reduction shall be added to or deducted from the Contract Price as the case may be.

Labour, materials and transport

6.2 If, by reason of any rise or fall in the Cost of labour or in the Cost of material or transport above or below such Cost ruling at the date of the Tender, the Cost to the Contractor of performing his obligations under the Contract shall be increased or reduced, the amount of such increase or reduction shall be added to or deducted from the Contract Price as the case may be. No account shall be taken of any amount by which any such Cost incurred by the Contractor has been increased by the default or negligence of the Contractor.

This sub-clause shall apply unless specifically excluded by the Special Conditions. Where this sub-clause does apply then unless otherwise agreed, increases or decreases shall be calculated and determined by reference to a formula to be specified in the Special Conditions.

Agreement

Agreement

7.1 Either party shall be entitled to require the other to enter into an agreement in the form annexed with such modifications as may be necessary within 45 days after the Letter of Acceptance. The expenses of preparing, completing and stamping the agreement shall be borne by the party making such request and he shall provide the other party free of charge with a copy of the agreement.

Performance bond or guarantee

Provision of bond or guarantee

8.1 If required by the Purchaser the Contractor shall provide the bond or guarantee of an insurance company, a bank or other surety for the due performance of the Contract. Unless otherwise specified in the Special Conditions, the terms of the bond or guarantee shall be in the form annexed to these general conditions.

Unless otherwise specified in the Contract the Contractor shall provide the bond or guarantee at his own expense.

The amount of the bond, the period of its validity, the procedure to be followed for its forfeiture, the arrangements for its release and the currency of any monetary transactions involved shall be stated in the Special Conditions.

Failure to provide bond or guarantee

8.2 If the Contractor shall have failed to provide the bond or guarantee within 30 days after the date of the Letter of Acceptance or within such further period as may be advised by the Purchaser, the Purchaser shall be entitled to terminate the Contract by seven days' notice to the Contractor. In the event of termination under this sub-clause the Contractor shall have no liability to the Purchaser other than to repay to the Purchaser all Cost properly incurred by the Purchaser incidental to the obtaining of new tenders.

Details confidential

Details confidential

9.1 The Purchaser and the Contractor shall treat the details of the Contract and any information made available in relation thereto as private and confidential and neither of them shall publish or disclose the same or any particulars thereof (save in so far as may be necessary for the purposes of the Contract) without the previous consent of the other, provided that nothing in this clause shall prevent the publication or disclosure of any such information that has come within the public domain otherwise than by breach of this clause.

Notices

Notices to purchaser and engineer

10.1 Any notice to be given to the Purchaser or to the Engineer under the Contract shall be served by sending the same by post, telex, cable or facsimile transmission to, or by leaving the same at, the respective addresses nominated for that purpose in the Special Conditions.

Notices to contractor

10.2 All certificates, notices or decisions, instructions and orders to be given by the Engineer or the Purchaser under the Contract shall be served by sending the same by post, telex, cable or facsimile transmission to, or by leaving the same at, the Contractor's principal place of business or such other address as the Contractor shall nominate for that purpose.

Service of notices	10.3 Any notice sent by telex, cable or facsimile transmission shall be deemed to have been served at the time of transmission. A notice sent by post shall be deemed to have been served four days after posting.

Purchaser's general obligations

Access to site	11.1 The Purchaser shall give the Contractor access (but not exclusive access) to the Site on the date specified in the Contract. If no date is stated then access shall be given in reasonable time having regard to the Time for Completion. The Purchaser shall provide such roads and other means of access to the Site as may be stated in the Specification, subject to such limitations as to use as may be imposed.
Wayleaves, consents, etc.	11.2 The Purchaser shall, within the times stated in the Programme or, if not so stated, before the time specified for delivery of any Plant to the Site, obtain all consents, wayleaves and approvals in connection with the regulations and bylaws of any local or other authority which shall be applicable to the Works on the Site.
Import permits, licences and duties	11.3 The Purchaser shall obtain all import permits or licences required for any part of the Plant or Works within the time stated in the Programme or, if not so stated, in reasonable time having regard to the time for delivery of the Plant and the Time for Completion. The Purchaser shall pay or reimburse to the Contractor all customs and import duties arising upon the importation of Plant into the country in which the Plant is to be erected. In the event that the Purchaser shall fail to obtain such import permits or licences then the additional Cost reasonably incurred by the Contractor in consequence thereof shall be added to the Contract Price.
Foundations, etc.	11.4 Buildings, structures, foundations, approaches or work, equipment or materials to be provided by the Purchaser shall be provided within the time specified in the Contract or in the Programme, shall be of the quality specified and in a condition suitable for the efficient transport, reception, installation and maintenance of the Works.
Purchaser's lifting equipment	11.5 The Purchaser shall at the Contractor's request and expense operate for the purposes of the Works any suitable lifting equipment belonging to the Purchaser that may be available on the Site and of which details are given in the Special Conditions. The Purchaser shall during such operation retain control of and be responsible for the safe working of the lifting equipment.
Utilities and power	11.6 The Purchaser shall make available on the Site for use by the Contractor for the purposes of the Works such supplies of electricity, water, gas, air and other services as may be specified in the Special Conditions. Such supplies shall be made available at the point(s) specified in the Contract on the terms mentioned in sub-clause 18.2 (Site services).
Power, etc. for tests on site	11.7 Where the Contract provides for any tests on Site, the Purchaser shall, unless otherwise stated in the Special Conditions, provide free of charge such fuel, electricity, skilled and unskilled labour, materials, stores, water, apparatus, instruments and feedstocks as may be requisite and as may reasonably be requested by the Contractor to enable the tests to be carried out effectively.
Breach of purchaser's general obligations	11.8 In the event that the Purchaser shall be in breach of any of his general obligations imposed by this clause then the additional Cost reasonably incurred by the Contractor in consequence thereof shall be added to the Contract Price.
Assistance with laws and regulations	12.1 Where the Works are to be erected outside the Contractor's country the Purchaser shall assist the Contractor to ascertain the nature and extent of and to comply with any laws, regulations, orders or bylaws of any local or national authority having the force of law in the country where the Works are to be

Contractor's obligations

Contractor's general obligations

13.1 The Contractor shall, subject to the provisions of the Contract, with due care and diligence, design, manufacture, deliver to Site, erect and test the Plant, execute the Works and carry out the Tests on Completion within the Time for Completion. The Contractor shall also make good defects in the Works and provide specialist advice to enable the Performance Tests (where these are included) to be carried out by the Purchaser. The Contractor shall, subject to sub-clause 11.7 (Power, etc. for tests on site), provide all labour, skilled and unskilled, the supervision thereof and all Contractor's Equipment required for the execution of the Works.

Manner of execution

13.2 The Works shall be manufactured and executed in the manner set out in the Specification or, where not so set out, to the reasonable satisfaction of the Engineer and all work on Site shall be carried out in accordance with such reasonable directions as the Engineer may give.

Contractor's design

13.3 The Contractor shall be responsible for the detailed design of the Plant and of the Works in accordance with the requirements of the Specification. In so far as the Contractor is required by the Contract or is instructed by the Engineer to comply with any detailed design provided by the Purchaser or the Engineer the Contractor shall be responsible for such design unless within a reasonable time after receipt thereof he shall have given notice to the Engineer disclaiming such responsibility.

Unless otherwise provided in the Contract the Contractor does not warrant that the Works as described in the Specification or the incorporation thereof within some larger project will satisfy the Purchaser's requirements.

Programme

14.1 Within the time stated in the Contract or, if no time is stated, within 30 days after the Letter of Acceptance, the Contractor shall submit to the Engineer for his approval the Programme for the execution of the Works showing:

(a) the sequence and timing of the activities by which the Contractor proposes to carry out the Works (including design, manufacture, delivery to Site, erection and testing),

(b) the anticipated numbers of skilled and unskilled labour and supervisory staff required for the various activities when the Contractor is working on Site,

(c) the respective times for submission by the Contractor of drawings and operating and maintenance instructions and for the approval thereof by the Engineer,

(d) the times by which the Contractor requires the Purchaser:

    (i) to furnish any drawings or information,

    (ii) to provide access to Site,

    (iii) to have completed any necessary civil engineering or building work (including foundations for the Plant),

    (iv) to have obtained any import permits or licences, consents, wayleaves and approvals necessary for the purposes of the Works,

    (v) to provide electricity, water, gas, air and other services on the Site or any equipment, materials or services which are to be provided by the Purchaser.

Copyright © 2000 Institution of Electrical Engineers

Form of programme

14.2 The Programme shall be in such form as may be specified in the Special Conditions or, if not so specified, as may reasonably be required by the Engineer.

Approval of programme

14.3 Approval by the Engineer of the Programme shall not relieve the Contractor of any of his obligations under the Contract.

Alterations to programme

14.4 The Contractor shall not without the Engineer's consent make any material alteration to the approved Programme.

Revision of programme

14.5 If the Engineer decides that progress under the Contract does not match the Programme, he may order the Contractor to revise the Programme. The Contractor shall thereafter revise the Programme to show the modifications necessary to ensure completion of the Works within the Time for Completion.

If modifications are required for reasons for which the Contractor is not responsible, the Cost of producing the revised Programme shall be added to the Contract Price.

Rate of progress

14.6 The Engineer shall notify the Contractor if the Engineer decides that the rate of progress of the Works or of any Section is too slow to meet the Time for Completion and that this is not due to a circumstance for which the Contractor is entitled to an extension of time under sub-clause 33.1 (Extension of time for completion).

Following receipt of such a notice the Contractor shall take such steps as may be necessary and as the Engineer may approve to remedy or mitigate the likely delay, including revision of the Programme. The Contractor shall not be entitled to any additional payment for taking such steps.

Drawings

15.1 The Contractor shall submit to the Engineer for approval:

(a) within the times given in the Contract or in the Programme such drawings, samples, patterns, models or information (including calculations) as may be called for therein, and in the numbers therein required;

(b) during the progress of the Works within such reasonable times as the Engineer may require such drawings of the general arrangement and details of the Works as may be specified in the Contract or as the Engineer may reasonably require.

The Engineer shall signify his approval or disapproval thereof. If he fails to do so within the time given in the Contract or the Programme, or, if no time limit is specified, within 30 days of receipt, they shall be deemed to have been approved.

Approved drawings, samples, patterns and models shall be signed or otherwise identified by the Engineer.

The Contractor shall supply additional copies of approved drawings in the form and numbers stated in the Contract.

Consequences of disapproval of drawings

15.2 Any drawings, samples, patterns or models which the Engineer disapproves shall be modified and re-submitted without delay.

Approved drawings

15.3 Approved drawings shall not be departed from except as provided in clause 27 (Variations).

Inspection of drawings

15.4 The Engineer shall have the right at all reasonable times to inspect all drawings of any part of the Works.

Foundation, etc. drawings

15.5 The Contractor shall provide within the times stated in the Contract or in the Programme, drawings showing how the Plant is to be affixed and any other information required for

- preparing suitable foundations or other means of support,
- providing suitable access for the Plant and any necessary equipment to the point on Site where the Plant is to be erected, and
- making necessary connections to the Plant.

Operating and maintenance instructions

15.6 Within the time or times stated in the Contract or in the Programme the Contractor shall supply operating and maintenance instructions and drawings of the Works as built. These shall be in such detail as will enable the Purchaser to operate, maintain, dismantle, reassemble and adjust all parts of the Works.

Instructions and drawings shall be supplied in the form and numbers stated in the Contract.

The Works shall not be considered as having been delivered for the purposes of taking-over until such instructions and drawings have been supplied to the Purchaser.

Purchaser's use of drawings, etc. supplied by contractor

15.7 Drawings and information supplied by the Contractor may be used by the Purchaser only for the purposes of completing, maintaining, adjusting and repairing the Works. No licence is granted to the Purchaser to copy or use drawings or information so supplied in order to make or have made spare parts for the Works. Drawings or information supplied by the Contractor shall not without the Contractor's consent be used, copied or communicated to a third party by the Engineer or the Purchaser otherwise than as strictly necessary for the purposes of the Contract.

Contractor's use of drawings, etc. supplied by purchaser or engineer

15.8 Drawings and information supplied by the Purchaser and the Engineer to the Contractor for the purposes of the Tender and the Contract shall remain the property of the Purchaser. They shall not without the consent of the Purchaser be used, copied or communicated to a third party by the Contractor otherwise than as strictly necessary for the purposes of the Contract.

Manufacturing drawings, etc.

15.9 Notwithstanding any other provisions of the Contract the Contractor shall not be required to provide to the Purchaser or to the Engineer shop drawings nor any Contractor's confidential manufacturing drawings, designs, or know-how nor the confidential details of manufacturing practices, processes or operations.

Errors in drawings, etc. supplied by contractor

16.1 Notwithstanding approval by the Engineer of drawings, samples, patterns, models or information submitted by the Contractor, the Contractor shall be responsible for any errors, omissions or discrepancies therein unless they are due to incorrect drawings, samples, patterns, models or information supplied by the Purchaser or the Engineer.

The Contractor shall bear any costs he may incur as a result of delay in providing such drawings, samples, patterns, models or information or as a result of errors, omissions or discrepancies therein, for which the Contractor is responsible.

The Contractor shall at his own expense carry out, or bear the reasonable cost of, any alterations or remedial work necessitated by such errors, omissions or discrepancies for which he is responsible and modify the drawings, samples, patterns, models or information accordingly.

The performance of his obligations under this clause shall be in full satisfaction of the Contractor's liability under this clause but shall not relieve him of his liability under clause 34 (Delay).

Errors in drawings, etc. supplied by purchaser or engineer

16.2 The Purchaser shall be responsible for errors, omissions or discrepancies in drawings and written information supplied by him or by the Engineer. The Purchaser shall at his own expense carry out any alterations or remedial work necessitated by such errors, omissions or discrepancies for which he is responsible or pay the Contractor the Cost incurred by the Contractor in carrying out in accordance with the Engineer's instructions any such alterations or remedial work so necessitated.

Contractor's representatives and workmen

17.1 The Contractor shall employ one or more competent representatives, whose name or names shall have been notified previously to the Engineer by the Contractor, to superintend the carrying out of the Works on the Site. The said representative, or if more than one shall be employed, then one of such representatives, shall be present on the Site during working hours, and any orders or instructions which the Engineer may give to the said representative of the Contractor shall be deemed to have been given to the Contractor.

Objection to representatives

17.2 The Engineer shall be entitled by notice to the Contractor to object to any representative or person employed by the Contractor in the execution of or otherwise about the Works who shall, in the opinion of the Engineer, misconduct himself or be incompetent or negligent, and the Contractor shall remove such person from the Works.

Returns of labour

17.3 The Contractor shall, if required by the Engineer, send to the Engineer's Representative a detailed return of the supervisory staff and the numbers of labour in the categories from time to time employed by the Contractor and his Sub-Contractors upon the Site. The returns shall be provided in such form and with such frequency as the Engineer may reasonably require.

Fencing, guarding, lighting and watching

18.1 The Contractor shall be responsible for the proper fencing, guarding, lighting and watching of all the Works on the Site until taken over and for the proper provision of temporary roadways, footways, guards and fences as far as may be necessary for the Works and for the accommodation and protection of the owners and occupiers of adjacent property, the public and others. The Contractor shall not use any naked light on the Site without the specific consent of the Engineer or the Engineer's Representative.

Site services

18.2 The Contractor shall provide any apparatus necessary for the use of such supplies of electricity, water, gas, air and other services as are made available for the Contractor's use by the Purchaser under sub-clause 11.6 (Utilities and power). The Contractor shall pay the Purchaser for such use at the rates specified in the Special Conditions or, if not so specified, at such reasonable rates as the Engineer shall determine.

Clearance of site

18.3 From time to time during the progress of the Works the Contractor shall clear away and remove from the Site all surplus materials and rubbish and, on completion, all Contractor's Equipment. The Contractor shall at all times leave the Site and the Works clean and in a safe and workmanlike condition to the Engineer's reasonable satisfaction.

Opportunities for other contractors

18.4 The Contractor shall, in accordance with the Engineer's requirements, afford all reasonable opportunities for carrying out their work to any other contractors employed by the Purchaser and to the workmen of the Purchaser who may be employed in the execution on or near the Site of any work not included in the Contract or of any contract which the Purchaser may enter into in connection with or ancillary to the Works. If, however, the Contractor shall, on the written request of the Engineer or the Engineer's Representative, make available to any such other contractor or to the Purchaser any Contractor's Equipment or provide any other service of whatsoever nature, the Purchaser shall pay to the Contractor in respect of such use or service such sum or sums as shall, in the opinion of the Engineer, be reasonable.

Hours of work

19.1 Unless otherwise provided in the Contract the Purchaser shall give the Contractor facilities for carrying out the Works on the Site continuously during the normal working hours generally recognised in the district. The Engineer may, after consulting with the Contractor, direct that work shall be done at

Copyright © 2000 Institution of Electrical Engineers

other times if it shall be practicable in the circumstances for work to be so done. The extra Cost of work so done shall be added to the Contract Price unless such work has, by the default of the Contractor, become necessary to ensure the completion of the Works within the Time for Completion.

No night or rest day working

19.2 No work shall be carried out on Site during the night or on locally recognised days of rest without the consent of the Engineer or the Engineer's Representative unless the work is unavoidable or necessary for the protection of life or property or for the safety of the Works, in which case the Contractor shall immediately advise the Engineer or the Engineer's Representative. The Engineer or the Engineer's Representative shall not withhold any such consent if work at night or on rest days is considered by the Contractor to be necessary to meet the Time for Completion.

Safety

20.1 The Contractor shall be responsible for the adequacy, stability and safety of his operations on Site and shall comply with the Purchaser's safety regulations applicable at the Site unless specifically authorised by the Engineer to depart therefrom in any particular circumstances.

Extraordinary traffic

21.1 The Contractor shall use every reasonable means to prevent damage to any of the highways or bridges on the routes to the Site by any traffic of the Contractor or any of his Sub-Contractors.

Special loads

21.2 Should the Contractor consider that the moving of one or more loads of Plant or Contractor's Equipment is likely to damage any highway or bridge unless special protection or strengthening is carried out, then the Contractor shall before moving the load notify the Engineer. The Contractor shall in the notice state the weight and other particulars of the load to be moved and his proposal for protecting or strengthening the highway or bridge.

Unless within fourteen days of receipt of such notice the Engineer by notice directs that such protection or strengthening is unnecessary, then the Contractor shall carry out his proposals with any modification thereof that the Engineer may require.

If there are items in the rates specified in the schedule of prices applicable to such work the Contractor shall be paid for such work at the rates specified. If there are no such rates, the Cost of the work shall be certified by the Engineer and added to the Contract Price.

Extraordinary traffic claims

21.3 If the Contractor shall receive any claim in respect of damage or injury to highways or bridges arising out of the execution of the Works, he shall immediately report the claim to the Engineer. The Purchaser shall then negotiate the settlement of and pay all sums due in respect of such claim. The Purchaser shall indemnify the Contractor in respect of the claim and in respect of all proceedings, damages, Cost, charges and expenses in relation thereto.

If the Engineer decides that any part of such claim results from the negligence of the Contractor or from the Contractor's failure to perform his obligations under sub-clauses 21.1 (Extraordinary traffic) and 21.2 (Special loads) then the Engineer shall certify the amount thereof which shall be deducted from the Contract Price.

Waterborne traffic

21.4 In the event that the Contractor uses waterborne transport the foregoing provisions of this clause shall be construed as though the word 'highway' included a lock, dock, sea wall or other structure relating to a waterway.

Setting out

22.1 The Contractor shall accurately set out the Works in relation to original points, lines and levels of reference given by the Engineer in writing and provide all necessary instruments, appliances and labour therefor.

If, at any time during the execution of the Works, any error appears in the positions, levels, dimensions or alignment of the Works, the Contractor shall rectify the error.

The Contractor shall bear the Cost of rectifying the error, unless the error results from incorrect information supplied in writing by the Purchaser, the Engineer or the Engineer's Representative, or from default by another contractor, in which case the Cost incurred by the Contractor shall be added to the Contract Price.

The Contractor shall identify, protect and preserve bench marks, sight rails, pegs and other things used in setting out the Works.

Inspection and testing of plant before delivery

Engineer's entitlement to test

23.1 The Engineer shall be entitled at all reasonable times during manufacture to inspect, examine and test on the Contractor's premises or elsewhere the materials and workmanship and performances of all Plant. If any part of the Plant is being manufactured on premises, other than the Contractor's own, the Contractor shall obtain permission for the Engineer to inspect, examine and test as if the Plant were being manufactured on the Contractor's premises. Such inspection, examination and testing shall not release the Contractor from any obligation under the Contract.

Date for test or inspection

23.2 The Contractor shall agree with the Engineer the date on and the place at which any Plant will be ready for testing or inspection as provided in the Contract. The Engineer shall give the Contractor 24 hours' notice of his intention to attend the test or inspection. If the Engineer shall not attend at the place so named on the date agreed, the Contractor may proceed with the test or inspection which shall be deemed to have been made in the Engineer's presence. The Contractor shall forthwith forward to the Engineer duly certified copies of the results of the test or inspection.

Services for test or inspection

23.3 The Contractor shall provide free of charge such assistance, labour, materials, electricity, fuel, stores, apparatus, instruments and other things as may be requisite and as may be reasonably demanded to carry out the test or inspection.

Certificate of test or inspection

23.4 When the Engineer is satisfied that any Plant has passed the test or inspection referred to in this clause he shall forthwith issue to the Contractor a certificate to that effect.

Failure on test or inspection

23.5 If after inspecting, examining or testing any Plant the Engineer shall decide that such Plant or any part thereof is defective or not in accordance with the Contract, he may reject the said Plant or part thereof by giving to the Contractor within 14 days notice of such rejection, stating therein the grounds upon which the said decision is based. Following any such rejection the Contractor shall make good or otherwise repair or replace the rejected Plant and resubmit the same for test or inspection in accordance with this clause and all expenses reasonably incurred by the Purchaser in attending or in consequence of such re-testing or inspection and the Engineer's attendance shall be deducted from the Contract Price.

Delivery

24.1 The Contractor shall apply in writing to the Engineer for permission to deliver any Plant or Contractor's Equipment to the Site. No Plant or Contractor's Equipment may be delivered to the Site without the Engineer's written permission.

The Contractor shall be responsible for the reception and unloading on Site of all Plant and Contractor's Equipment delivered for the purposes of the Contract.

Suspension of work, delivery or erection

Instructions to suspend

25.1 The Engineer may at any time instruct the Contractor to suspend the progress of the whole or any part of the Works.

If by reason of any delay or failure on the part of the Engineer, or of failure by the Engineer to give the permission referred to in clause 24.1 (Delivery), or from any cause for which the Purchaser or some other contractor employed by him is responsible, the Contractor is prevented from either:

(a) delivering to the Site any Plant which is ready for delivery at the time for delivery thereof specified in the Programme, or if no time is specified, at the time appropriate for it to be delivered having regard to the Time for Completion ('the Normal Delivery Date'), or

(b) erecting any Plant which has been delivered to the Site

then the Engineer shall be deemed to have given instructions to suspend the progress of the Works to the extent that progress is dependent on the delivery or erection of such Plant.

The Contractor shall during suspension store, preserve, protect and otherwise secure the Works and/or Plant affected and insure the same to the extent required by the Engineer.

Unless otherwise instructed by the Engineer, the Contractor shall during any suspension affecting the progress of the Works on Site maintain his staff, labour and Contractor's Equipment on or near the Site ready to proceed with the Works upon receipt of the Engineer's further instructions.

Additional cost caused by suspension

25.2 Any additional Cost incurred by the Contractor in complying with the provisions of and/or the Engineer's instructions under sub-clause 25.1 (Instructions to suspend) shall be added to the Contract Price.

Payment for plant affected by suspension

25.3 The Contractor shall be entitled to payment for Plant, work on which, or the delivery of which at the Normal Delivery Date, has been suspended for more than 30 days and the Contract Value of such Plant as at the date of suspension shall be included in an interim certificate of payment under sub-clause 39.3 (Issue of payment certificate).

The Contractor shall not be entitled to a certificate of payment until he has sufficiently marked the Plant as the Purchaser's property under sub-clause 37.2 (Marking of plant) and has insured it in accordance with sub-clause 47.1 (Insurance of works) as if the Plant were on Site.

Disallowance of additional cost or payment

25.4 The Contractor shall not be entitled to be paid any additional Cost under sub-clause 25.2 (Additional cost caused by suspension) nor to any payment under sub-clause 25.3 (Payment for plant affected by suspension) if suspension is necessary by reason of default on the part of the Contractor or for the proper execution or the safety of the Works or Plant, save where such necessity results from any act or default of the Engineer or the Purchaser or the occurrence of any of the Purchaser's Risks.

Resumption of work, delivery or erection

25.5 At any time after suspension under sub-clause 25.1 (Instructions to suspend) the Engineer may give notice to the Contractor to proceed with the delivery or erection of Plant and/or work the subject of suspension under this clause.

If suspension has continued for more than 90 days and the suspension is not necessitated by any of the reasons stated in sub-clause 25.4 (Disallowance of additional cost or payment) the Contractor may by notice to the Engineer require him to give notice to proceed within 30 days.

If notice to proceed is not given within that time the Contractor may elect to treat the suspension as an omission under clause 27 (Variations) of the part of the Works affected thereby. If the suspension affects the whole of the Works the Contractor may terminate the Contract in which event he shall be entitled to be paid in accordance with sub-clause 51.2 (Payment on termination due to purchaser's default) as if the Contract had been terminated under sub-clause 51.1 (Notice of termination due to purchaser's default). If the Contractor does not elect to treat the suspension as an omission or to terminate the Contract, as the case may be, he shall be entitled to be paid the Contract Value of the Plant affected by the suspension.

Upon receipt of notice to proceed, the Contractor shall examine the Plant and work affected by the suspension. The Contractor shall make good any deterioration or defect in or loss of such Plant or work that may have occurred during suspension. The Cost incurred in making such examination and of making good and resuming work shall be added to the Contract Price, provided that the Contractor shall not be entitled to be paid any Cost incurred in making good any deterioration, defect or loss caused by defective materials or workmanship or by the Contractor's failure to comply with any instructions of the Engineer under sub-clause 25.1 (Instructions to suspend).

Effect of suspension on defects liability

25.6 If the Contractor shall, solely in consequence of suspension, be required to perform his obligations under clause 36 (Defects liability) in relation to defects in any Plant at a time which is more than three years after the Normal Delivery Date therefor, the additional Cost incurred by the Contractor shall be added to the Contract Price.

Defects before taking-over

Defects before taking-over

26.1 Without prejudice to the Purchaser's rights under sub-clause 23.5 (Failure on test or inspection) if, in respect of any part of the Works not yet taken over, the Engineer shall at any time:

(a) decide that any work done or Plant supplied or materials used by the Contractor or any Sub-Contractor is or are defective or not in accordance with the Contract, or that such part is defective or does not fulfil the requirements of the Contract (all such matters being hereinafter in this clause called 'defects'), and

(b) as soon as reasonably practicable notify the Contractor of the said decision, specifying particulars of the defects alleged and

(c) so far as may be necessary place the Plant at the Contractor's disposal,

then the Contractor shall with all speed and, except as provided in sub-clause 25.5 (Resumption of work, delivery or erection) at his own expense, make good the defects so specified. If the Contractor fails so to do the Purchaser may, provided he does so without undue delay, take, at the expense of the Contractor, such steps as may in all the circumstances be reasonable to make good such defects. All work done, or Plant provided by the Purchaser to replace defective Plant, shall comply with the Contract and shall be obtained at reasonable prices and where reasonably practicable under competitive conditions. The Contractor shall be entitled at his own expense to remove and retain all Plant that the Purchaser may have replaced. Nothing contained in this clause shall affect any claim by the Purchaser under clause 34 (Delay).

Variations

Meaning of variation

27.1 In the Conditions the term 'variation' means any alteration of the Works whether by way of addition, modification or omission.

Engineer's power to vary

27.2 The Engineer alone shall have the power until the whole of the Works have been taken over under clause 29 (Taking-over) to instruct the Contractor by notice to make any variations to the Works.

The Contractor shall carry out such variations and be bound by the Conditions in so doing as though the variations were stated in the Specification.

As soon as possible after having received any such instruction, the Contractor shall notify the Engineer if, in the Contractor's opinion, the variation will involve an addition to or deduction from the Contract Price.

No such variation shall, together with any variations already ordered, involve a net addition to or deduction from the Contract Price of more than 15 per cent thereof unless the Contractor and the Purchaser consent thereto in writing.

Nothing in this sub-clause shall prevent the Contractor from making proposals to the Engineer for variations to the Works, but no variation so proposed shall be carried out by the Contractor except as directed in writing by the Engineer.

Valuation of variations

27.3 The amount to be added to or deducted from the Contract Price shall, if not the subject of a quotation from the Contractor which has been accepted by the Purchaser prior to the variation having been ordered, be determined by the Engineer in accordance with the rates specified in the schedules of prices, if applicable. Where rates are not contained in the said schedules or are not applicable then the amount shall be such sum as is in all the circumstances reasonable. Due account shall be taken of any partial execution of the Works which is rendered useless by any such variation.

Contractor's records of costs

27.4 In any case where the Contractor is instructed to proceed with a variation prior to the determination of the value thereof under sub-clause 27.3 (Valuation of variations) the Contractor shall keep contemporary records of the Cost of making the variation and of time expended thereon. Such records shall be open to inspection by the Engineer at all reasonable times.

Notice and confirmation of variations

27.5 When ordering any variation to any part of the Works, the Engineer shall give the Contractor such reasonable notice as will enable him to make his arrangements accordingly.

In cases where Plant is already manufactured, or in the course of manufacture, or any work done or drawings or patterns made that require to be altered, the Contractor shall be entitled to be paid the Cost of such alterations.

If, in the opinion of the Contractor, any such variation is likely to prevent or prejudice him from or in fulfilling any of his obligations under the Contract, he shall notify the Engineer thereof with full supporting details. The Engineer shall decide forthwith whether or not the variation shall be carried out.

If the Engineer confirms his instructions in writing the said obligations shall be modified to such an extent as may be justified. Until the Engineer so confirms his instructions, they shall be deemed not to have been given.

Progress with variations

27.6 The Contractor shall on receipt of the Engineer's instructions under sub-clause 27.2 (Engineer's power to vary), or confirmation of instructions under sub-clause 27.5 (Notice and confirmation of variations), immediately proceed to carry out such instructions, unless the Contractor has notified the Engineer that the variation in his opinion will involve a net addition to or deduction from the Contract Price of more than 15 per cent.

Subject thereto, the carrying out of such instructions shall not, without the consent of the Engineer, be delayed pending agreement on price.

Tests on completion

Notice of tests

28.1 The Contractor shall give to the Engineer 21 days' notice of the date after which he will be ready to make the Tests on Completion. Unless otherwise agreed the Tests on Completion shall take place within 10 days after the said date on such day or days as the Engineer shall notify to the Contractor.

Time for tests

28.2 If the Engineer fails to appoint a time after having been asked so to do or to attend at any time or place duly appointed for making the Tests on Completion, the Contractor shall be entitled to proceed in his absence and the Tests on Completion shall be deemed to have been made in the presence of the Engineer. The Contractor shall forthwith forward to the Engineer duly certified copies of the results of the Tests on Completion.

Delayed tests

28.3 If the Tests on Completion are being unduly delayed by the Contractor, the Engineer may, by notice, call upon the Contractor to make them within 21 days from the receipt of the said notice. The Contractor shall make the Tests on Completion on such days within the said 21 days as the Contractor may fix and of which he shall give notice to the Engineer. If the Contractor fails to make the Tests on Completion within the time aforesaid the Purchaser or the Engineer may proceed therewith at the risk and expense of the Contractor and the Cost thereof shall be deducted from the Contract Price. If the Contractor shall establish that the Tests on Completion were not being unduly delayed, the Tests on Completion so made shall be at the risk and expense of the Purchaser.

Repeat tests

28.4 If any part of the Works fails to pass the Tests on Completion they shall be repeated within a reasonable time upon the same terms and conditions. All Cost which the Purchaser may incur in the repetition of the Tests on Completion shall be deducted from the Contract Price. The provisions of this sub-clause shall also apply to any tests carried out under sub-clause 36.7 (Further tests).

Consequences of failure to pass tests on completion

28.5 If the Works or any Section fails to pass the Tests on Completion (including any repetition thereof) the Contractor shall take whatever steps may be necessary to enable the Works or the Section to pass the Tests on Completion and shall thereafter repeat them, unless any time limit specified in the Contract for the passing thereof shall have expired, in which case the Engineer shall be entitled to reject the Works or the Section and the Purchaser shall be entitled to proceed in accordance with clause 49 (Contractor's default).

Taking-over

Taking-over by sections

29.1 If the Contract provides for the Works to be taken over by Sections the provisions of this clause shall apply to each such Section as it applies to the Works.

Taking-over certificate

29.2 When the Works have passed the Tests on Completion and are complete (except in minor respects that do not affect their use for the purpose for which they are intended) the Engineer shall issue a certificate to the Contractor and to the Purchaser (herein called a 'Taking-Over Certificate'). The Engineer shall in the Taking-Over Certificate certify the date upon which the Works passed the Tests on Completion and were so complete.

The Purchaser shall be deemed to have taken over the Works on the date so certified. Except as permitted by clause 30 (Use before taking-over) the Purchaser shall not use the Works before they are taken over.

Effect of taking-over certificate

29.3 With effect from the date of taking-over as stated in the Taking-Over Certificate, risk of loss or damage to the Works or to the Section to which the Taking-Over Certificate relates (other than any parts thereof excluded by the terms of the Taking-Over Certificate) shall pass to the Purchaser and he shall take possession thereof.

Outstanding work

29.4 The Contractor shall rectify or complete to the reasonable satisfaction of the Engineer within the time stated in the Taking-Over Certificate any outstanding items of work or Plant noted as requiring rectification or as

Copyright © 2000 Institution of Electrical Engineers

incomplete. In the event the Contractor fails to do so, the Purchaser may arrange for the outstanding work to be done and the Cost thereof shall be certified by the Engineer and deducted from the Contract Price.

Use before taking-over

30.1 If, by reason of any default on the part of the Contractor, a Taking-Over Certificate has not been issued in respect of the whole of the Works within one month after the Time for Completion, the Purchaser shall be entitled to use any Section or part of the Works in respect of which a Taking-Over Certificate has not been issued, provided the same is reasonably capable of being used. The Contractor shall be afforded the earliest possible opportunity of taking such steps as may be necessary to permit the issue of the Taking-Over Certificate. The provisions of sub-clause 43.1 (Care of the works) shall not apply to any Section or part of the Works while being so used by the Purchaser and clause 36 (Defects liability) shall apply thereto as if a Taking-Over Certificate had been issued from the date on which the Section or part was taken into use.

Interference with tests

31.1 If by reason of any act or omission of the Purchaser, the Engineer or some other contractor employed by the Purchaser, the Contractor shall be prevented from carrying out the Tests on Completion in accordance with clause 28 (Tests on completion) then, unless in the meantime the Works have been proved not to be substantially in accordance with the Contract, the Purchaser shall be deemed to have taken over the Works and the Engineer shall, upon the application of the Contractor, issue a Taking-Over Certificate accordingly.

Tests during defects liability period

31.2 In any case where a Taking-Over Certificate has been issued under sub-clause 31.1 (Interference with tests) the Contractor shall be under an obligation to carry out the Tests on Completion during the Defects Liability Period as and when required by 14 days' notice from the Engineer. Such allowances shall be made from the results required to be attained in the Tests on Completion as may be reasonable having regard to any use of the Works by the Purchaser prior to the Tests on Completion and to any deterioration therein which may have occurred since the issue of the Taking-Over Certificate in respect thereof. The additional Cost incurred by the Contractor in making the Tests on Completion in accordance with this sub-clause shall be certified by the Engineer and added to the Contract Price.

Time for completion

Time for completion

32.1 Subject to any requirement under the Contract for the completion of any Section before the completion of the whole of the Works, the Contractor shall so execute the Works that they shall be complete and pass the Tests on Completion (but not the Performance Tests, if any be included) within the Time for Completion.

Extension of time for completion

33.1 If, by reason of any variation ordered pursuant to clause 27 (Variations) or of any act or omission on the part of the Purchaser or the Engineer or of any industrial dispute or by reason of circumstances beyond the reasonable control of the Contractor arising after the acceptance of the Tender, the Contractor shall have been delayed in the completion of the Works, whether such delay occurs before or after the Time for Completion, then provided that the Contractor shall as soon as reasonably practicable have given to the Purchaser or the Engineer notice of his claim for an extension of time with full supporting details, the Engineer shall on receipt of such notice grant the Contractor from time to time in writing either prospectively or retrospectively such extension of the Time for Completion as may be reasonable.

Delays by sub-contractors

33.2 Any delay on the part of a Sub-Contractor which prevents the Contractor from completing the Works within the Time for Completion shall entitle the Contractor to an extension thereof provided such delay is due to a cause for which the Contractor himself would have been entitled to an extension of time under sub- clause 33.1 (Extension of time for completion).

Mitigation of consequences of delay

33.3 In all cases where the Contractor has given notice under sub-clause 33.1 (Extension of time for completion) the Contractor shall consult with the Engineer in order to determine the steps (if any) which can be taken to overcome or minimise the actual or anticipated delay. The Contractor shall thereafter comply with all reasonable instructions which the Engineer shall give in order to overcome or minimise such delay. If compliance with any such instruction shall cause the Contractor to incur additional Cost and the Contractor is entitled to an extension of time, the amount thereof shall be added to the Contract Price.

Delay

Delay in completion

34.1 If the Contractor fails to complete the Works in accordance with the Contract, save as regards his obligations under clauses 35 (Performance tests) and 36 (defects liability), within the Time for Completion, or if no time be fixed, within a reasonable time, there shall be deducted from the Contract Price or paid to the Purchaser by the Contractor the percentage stated in the Appendix of the Contract Value of such parts of the Works as cannot in consequence of the said failure be put to the use intended for each week between the Time for Completion and the actual date of completion. The amount so deducted or paid shall not exceed the maximum percentage stated in the Appendix of the Contract Value of such parts of the Works, and such deduction or payment shall, subject to sub-clause 34.2 (Prolonged delay), be in full satisfaction of the Contractor's liability for the said failure.

Prolonged delay

34.2 If any part of the Works in respect of which the Purchaser has become entitled to the maximum amount provided under sub-clause 34.1 (Delay in completion) remains uncompleted, the Purchaser may by notice to the Contractor require him to complete. Such notice shall fix a final Time for Completion which shall be reasonable having regard to such delay as has already occurred and to the extent of the work required for completion. If for any reason, other than one for which the Purchaser or some other contractor employed by him is responsible, the Contractor fails to complete within such time, the Purchaser may by further notice to the Contractor elect either:

(a) to require the Contractor to complete, or

(b) to terminate the Contract in respect of such part of the Works,

and recover from the Contractor any loss suffered by the Purchaser by reason of the said failure up to an amount not exceeding the sum stated in the Appendix or, if no sum be stated, that part of the Contract Price that is properly apportionable to such part of the Works as cannot by reason of the Contractor's failure be put to the use intended.

Performance tests

Time for performance tests

35.1 Where Performance Tests are included in the Contract they shall be carried out as soon as is reasonably practicable and within a reasonable time after the Works, or the Section of the Works to which such tests relate, have been taken over by the Purchaser.

Procedures for performance tests

35.2 Performance Tests shall be carried out by the Purchaser or the Engineer on his behalf under the supervision of the Contractor and in accordance with the procedures and under the operating conditions specified in the Contract and in accordance with such other instructions as the Contractor may give in the course of carrying out such tests.

Cessation of performance tests

35.3 The Purchaser, or the Engineer on his behalf, or the Contractor shall be entitled to order the cessation of any Performance Test if damage to the Works or personal injury are likely to result from continuation.

Adjustments and modifications

35.4 If the Works or any Section thereof fails to pass any Performance Test (or repetition thereof) or if any Performance Test is stopped before its completion, such test shall, subject to sub-clause 35.5 (Postponement of adjustments and modifications), be repeated as soon as practicable thereafter. Any additional Cost incurred by the Purchaser solely by reason of the repetition of any Performance Test shall be deducted from the Contract Price. The Purchaser shall permit the Contractor to make adjustments and modifications to any part of the Works before the repetition of any Performance Test and shall, if required by the Contractor, shut down any part of the Works for such purpose and re-start it after the adjustments and modifications have been made. All such adjustments and modifications shall be made by the Contractor with all reasonable speed and at his own expense. The Contractor shall, if so required by the Engineer, submit to the Engineer for his approval details of the adjustments and modifications which he proposes to make.

Postponement of adjustments and modifications

35.5 If the Works or any Section thereof fails to pass any Performance Test (or repetition thereof) and the Contractor in consequence proposes to make any adjustments or modifications thereto, the Engineer may notify the Contractor that the Purchaser requires the carrying out of such adjustments or modifications to be postponed. In such event the Contractor shall remain liable to carry out the adjustments or modifications and a successful Performance Test within a reasonable time of being notified to do so by the Engineer. If however the Engineer fails to give any such notice within one year of the date of taking-over of the Works or Section thereof, the Contractor shall be relieved of any such obligation and the Works or Section thereof shall be deemed to have passed such Performance Test.

Time for completion of performance tests

35.6 If the Contract provides that the Performance Tests (or repetition thereof) shall be completed within a specified time the Purchaser shall be entitled to use the Works as he thinks fit from the expiry of such time.

Evaluation of results of performance tests

35.7 The results of Performance Tests shall be compiled and evaluated jointly by the Purchaser, or the Engineer on his behalf, and the Contractor in the manner detailed in the Contract. Any necessary adjustments to the results to take account of any previous use of the Works by the Purchaser, the measuring tolerances and any differences between the operating conditions under which the Performance Tests were conducted and those detailed in the Specification or performance test schedule shall be made in accordance with the provisions of the Specification or, if the Specification contains no such provisions, then in such manner as is fair and reasonable.

Consequences of failure to pass performance tests

35.8 If the Works or any Section fails to pass the Performance Tests (or any repetition thereof) within the period specified in the Special Conditions or, if no period is specified, within a reasonable time:

(a) where liquidated damages for failure to achieve any guaranteed performances have been specified in the Special Conditions and the results are within the stipulated acceptance limits the Contractor shall pay or allow to the Purchaser the liquidated damages so specified. Upon payment or allowance of such liquidated damages by the Contractor the Purchaser shall accept the Works.

(b) where such damages have been so specified but the results are outside the stipulated acceptance limits, or where liquidated damages have not been so specified, the Purchaser shall be entitled to accept the Works or the Section subject to such reasonable reduction in the Contract Price as may be agreed by the Purchaser and the Contractor or, in default of agreement, as may be determined by arbitration under clause 52 (Disputes and arbitration).

(c) where such failure of the Works or the Section would deprive the Purchaser of substantially the whole of the benefit thereof the Engineer shall be entitled to reject the Works or the Section and the Purchaser shall be entitled to proceed in accordance with clause 49 (Contractor's default).

Defects liability

Defects after taking-over

36.1 In the Conditions the expression 'Defects Liability Period' means the period stated in the Special Conditions as the Defects Liability Period or, if no such period is stated, 12 months, calculated from the date of taking-over of the Works under clause 29 (Taking-over). Where any Section or part of the Works is taken over separately the Defects Liability Period in relation thereto shall commence on the date of taking-over thereof.

Making good defects

36.2 The Contractor shall be responsible for making good by repair or replacement with all possible speed at his expense any defect in or damage to any part of the Works which may appear or occur during the Defects Liability Period and which arises either:

(a) from any defective materials, workmanship or design, or

(b) from any act or omission of the Contractor done or omitted during the said period.

The Contractor's obligations under this clause shall not apply to any defects in designs furnished or specified by the Purchaser or the Engineer in respect of which the Contractor has disclaimed responsibility in accordance with sub-clause 13.3 (Contractor's design), nor to any damage to any part of the Works in consequence thereof.

Notice of defects

36.3 If any such defect shall appear or damage occur the Purchaser or the Engineer shall forthwith inform the Contractor thereof stating in writing the nature of the defect or damage. The provisions of this clause shall apply to all repairs or replacements carried out by the Contractor to remedy defects and damage as if the said repairs or replacements had been taken over on the date they were completed; however the Defects Liability Period in respect thereof shall not extend beyond two years from the date of taking-over or such other period as may be stated in the Special Conditions.

Extension of defects liability

36.4 The Defects Liability Period shall be extended by a period equal to the period during which the Works (or that part thereof in which the defect or damage to which this clause applies has appeared or occurred) cannot be used by reason of that defect or damage.

Delay in remedying defects

36.5 If any such defect or damage be not remedied within a reasonable time, the Purchaser may proceed to do the work at the Contractor's risk and expense provided that he does so in a reasonable manner and notifies the Contractor of his intention so to do. The Cost reasonably incurred by the Purchaser shall be deducted from the Contract Price or be paid by the Contractor to the Purchaser.

Removal of defective work

36.6 The Contractor may with the consent of the Engineer remove from the Site any part of the Works which is defective or damaged, if the nature of the defect or damage is such that repairs cannot be expeditiously carried out on the Site.

Further tests

36.7 If the repairs or replacements are of such a character as may affect the operation of the Works or any part thereof, the Purchaser or the Engineer may within one month after such repair or replacement give to the Contractor notice requiring that further Tests on Completion or Performance Tests be made, in which case such tests shall be carried out as provided in clauses 28 (Tests on completion) or 35 (Performance tests) as the case may be.

Contractor to search

36.8 The Contractor shall, if required by the Engineer in writing, search for the cause of any defect, under the direction of the Engineer. Unless such defect shall be one which the Contractor is responsible for making good under sub-clause 36.2 (Making good defects) the Cost of the work carried out by the Contractor in searching as aforesaid shall be borne by the Purchaser and added to the Contract Price.

Limitation of liability for defects

36.9 The Contractor's liability under this clause shall be in lieu of any contract term implied by law as to the quality or fitness for any particular purpose or the workmanship of any part of the Works taken over under clause 29 (Taking-over) and, save as expressed in this clause 36, neither the Contractor nor his Sub-Contractors, their respective servants or agents shall be liable, whether in contract, in tort (including but not limited to negligence) or by reason of breach of statutory duty or otherwise, in respect of defects in or damage to such part, or for any damage or loss of whatsoever kind attributable to such defects or damage or any work done or service or advice rendered in connection therewith.

For the purposes of this sub-clause the Contractor contracts on his own behalf and on behalf of and as trustee for his Sub-Contractors, servants and agents. Nothing in this clause 36 shall affect the liability of the Contractor under the Conditions in respect of any part of the Works not yet taken over or his liability for death or personal injury caused by his wilful or negligent acts or omissions.

Latent defects

36.10 If any defect of the kind referred to in sub-clause 36.2 (Making good defects) shall appear in any part of the Works within a period of three years after the date of the taking-over of such part of the Works, the same shall be made good by the Contractor by repair or replacement at the Contractor's option provided that the defect was caused by the gross misconduct of the Contractor as defined below and would not have been disclosed by a reasonable examination prior to the expiry of the Defects Liability Period.

'Gross misconduct' does not comprise each and every lack of care or skill but means an act or omission on the part of the Contractor which implies either a failure to pay due regard to the serious consequences which a conscientious and responsible contractor would normally foresee as likely to ensue or a wilful disregard of such consequences.

Vesting of plant, and contractor's equipment

Ownership of plant

37.1 Plant to be supplied pursuant to the Contract shall become the property of the Purchaser at whichever is the earlier of the following times:

(a) when Plant is delivered pursuant to the Contract

(b) when the Contractor becomes entitled to have the value of the Plant in question included in an interim certificate of payment.

Marking of plant

37.2 Where, prior to delivery, the property in Plant passes to the Purchaser, the Contractor shall, so far as is practicable, set the Plant aside and mark it as the Purchaser's property in a manner reasonably required by the Engineer.

Until the Plant has been so set aside and marked the Contract Value of the Plant shall not be included in any interim certificate of payment to which the Contractor might otherwise be entitled.

The Contractor shall permit the Engineer at any time upon reasonable notice to inspect any Plant which has become the property of the Purchaser and shall grant the Engineer access for such purpose to the Contractor's premises or procure the grant to the Engineer of access for such purpose to any other premises where such Plant may be located.

Copyright © 2000 Institution of Electrical Engineers

All such Plant shall be in the care and possession of the Contractor solely for the purposes of the Contract and shall not be within the ownership or disposition of the Contractor.

No interim certificate of payment issued by the Engineer shall prejudice his right to reject Plant which is not in accordance with the Contract. Upon any such rejection the property in the rejected Plant shall immediately revert to the Contractor.

Contractor's equipment

38.1 The Contractor shall within 30 days after the Letter of Acceptance provide to the Engineer a list of the Contractor's Equipment that the Contractor intends to use on the Site.

Contractor's equipment on site

38.2 All Contractor's Equipment shall, when brought on to the Site, be deemed to be intended exclusively for the execution of the Works. The Contractor shall not thereafter remove the same or any part thereof from the Site without the consent of the Engineer, which shall not be withheld in the case of Contractor's Equipment not currently required for the execution of the Works on Site.

Loss or damage to contractor's equipment

38.3 The Contractor shall be liable for loss of or damage to any of the Contractor's Equipment which may occur otherwise than through the default of the Purchaser or those for whom he is responsible.

Maintenance of contractor's equipment

38.4 The Contractor shall be responsible for maintaining Contractor's Equipment on Site in safe working order.

Certificates and payment

Application for payment

39.1 Unless otherwise provided in the Special Conditions the Contractor may make application to the Engineer for interim certificates of payment in respect of:

- Plant in the course of manufacture,
- Plant delivered or shipped and en route to the Site,
- work executed on the Site,
- claims for additional payment in accordance with the Conditions,
- Plant affected by suspension under sub-clause 25.1 (Instructions to suspend).

Form of application

39.2 Applications for interim certificates of payment shall be in the form of an invoice accompanied:

(a) in the case of Plant in the course of manufacture, by such evidence of the value of the work done as may be specified in the Special Conditions;

(b) in the case of Plant delivered, shipped or en route to the Site, by such evidence of delivery or shipment and payment of freight and insurance, bills of lading or documents of title and by such other documents as may be specified in the Special Conditions;

(c) in the case of work executed on the Site, by such evidence of the value of the work done as may be specified in the Special Conditions.

(d) in the case of claims for additional payment, by the particulars required under sub-clause 41.1 (Notification of claims);

(e) under sub-clause 25.3 (Payment for plant affected by suspension), by such evidence of the value of the work done as the Engineer may reasonably require.

Copyright © 2000 Institution of Electrical Engineers

Issue of payment certificate

39.3 The Engineer shall issue an interim certificate of payment to the Contractor (with a copy to the Purchaser) within 14 days after receiving an application therefor which the Contractor was entitled to make.

Value included in certificates of payment

39.4 Every interim certificate of payment shall certify the total sum due to the Contractor from the Purchaser in accordance with the terms of payment specified in the Special Conditions in respect of:

– work done in the course of manufacture and/or duly executed on the Site and/or

– Plant shipped and/or Plant delivered to the Site including where appropriate freight, carriage and insurance charges and/or

– claims for additional payment and/or

– sub-clause 25.3 (Payment for plant affected by suspension)

up to the date named in the application for payment, less:

– the total of any sums previously certified in certificates of payment.

Provided that no sum shall be included in any interim certificate of payment in respect of any work or Plant which, in the reasonable opinion of the Engineer:

– does not comply with the Contract, or

– has been brought, and is at the date of the certificate, prematurely upon the Site.

Adjustments to certificates

39.5 If any sum shall become payable to the Contractor under the Contract otherwise than for work executed or Plant delivered, the amount thereof shall be included in the next certificate of payment.

If any sum shall become payable under the Contract by the Contractor to the Purchaser, whether by deduction from the Contract Price or otherwise, the amount thereof shall be deducted in the next certificate of payment.

Corrections to certificates

39.6 The Engineer may in any certificate of payment make any correction or modification that should properly be made in respect of any previous certificate.

Withholding certificate of payment

39.7 An interim certificate of payment shall not be withheld on account of defects of a minor character which are not such as to affect the use of the Works.

Effect of certificates of payment

39.8 No certificate of payment other than a final certificate of payment shall be relied upon as conclusive evidence of any matter stated therein, nor shall it affect or prejudice any right of the Purchaser or the Contractor against the other.

Application for final certificate of payment

39.9 The Contractor shall make application for the final certificate of payment forthwith after the Contractor's obligations under clause 36 (defects liability), other than under sub-clause 36.10 (Latent defects), have ceased and the Contractor has completed any outstanding remedial work thereunder. If a separate Taking-over Certificate has been issued in respect of a Section of the Works, the Contractor may apply for a final certificate of payment in respect thereof.

The application for the final certificate of payment shall be accompanied by a final account prepared by the Contractor in relation to those Sections of the Works to which the application relates. The final account shall give full details of the value of all Plant supplied and work done under the Contract and other sums certified by the Engineer, together with a detailed analysis and valuation of all claims to which the Contractor considers he is entitled under the Contract.

Value of final certificate of payment

39.10 A final certificate of payment shall certify the total amount payable to the Contractor under the Contract in respect of the Works or any Section thereof having regard to any addition to or deduction from the Contract Price provided for in the Conditions and claims in respect thereof made by the Contractor or the Purchaser, the total amounts paid on certificates of payment previously issued, and the balance payable whether by the Purchaser to the Contractor or by the Contractor to the Purchaser.

Issue of final certificate of payment

39.11 The Engineer shall issue to the Contractor (with a copy to the Purchaser) a final certificate of payment within 30 days after receiving an application therefor which the Contractor was entitled to make and which complies with all the requirements of sub-clause 39.9 (Application for final certificate of payment). For the purposes of this sub-clause time shall not start to run until the Contractor has provided to the Engineer all information in amplification of the final account that the Engineer may reasonably require.

Effect of final certificate of payment

39.12 A final certificate of payment shall be conclusive evidence:

– that the Works or Section to which such certificate relates is in accordance with the Contract;

– that the Contractor has performed all his obligations under the Contract in respect thereof; and

– of the value of the Works or Section.

Payment of the amount certified in a final certificate of payment shall be conclusive evidence that the Purchaser has performed all his obligations under the Contract in relation to the Works or Section thereof to which the certificate relates.

A final certificate of payment shall not be conclusive as to any matter dealt with in the certificate in the case of fraud or dishonesty relating to or affecting any such matter.

A final certificate of payment shall not be conclusive if any proceedings arising out of the Contract whether under clause 52 (Disputes and arbitration) or otherwise shall have been commenced by either party in relation to the Works or Section to which the certificate relates,

– before the final certificate of payment has been issued, or

– within three months thereafter.

No effect in the case of gross misconduct

39.13 Nothing in this clause shall affect the rights of the Purchaser or the obligations of the Contractor under sub-clause 36.10 (Latent defects).

Payment

40.1 The Purchaser shall pay to the Contractor the sum certified as due to the Contractor in a certificate of payment within 30 days after the date of issue thereof, unless otherwise specified in the Special Conditions.

Any payment made before delivery of Plant otherwise than a payment in respect of Plant in the course of manufacture or work done shall be subject to the Contractor first having furnished to the Purchaser a bond or guarantee of a bank or insurance company approved by the Purchaser, if so required by the Special Conditions.

Delayed payment

40.2 If payment of any sum payable under sub-clause 40.1 (Payment) is delayed, the Contractor shall be entitled to receive interest on the amount unpaid during the period of delay. The interest shall be at the rate of two per cent per annum above the average of the base rates of the London clearing banks in force from time to time during the period of delay, or at such other rate as may be specified in the Special Conditions. The Contractor shall be entitled to interest without formal notice and without prejudice to any other right or remedy.

Copyright © 2000 Institution of Electrical Engineers

Remedies on failure to certify or make payment

40.3 If the Engineer fails to issue a certificate of payment to which the Contractor is entitled or if the Purchaser fails to make any payment as provided in this clause subject to any deduction that the Purchaser is entitled to make under the Contract the Contractor shall be entitled:

(a) to stop work until the failure be remedied, by giving 14 days' notice to the Engineer and the Purchaser, in which event the additional Cost to the Contractor occasioned by the stoppage and the subsequent resumption of work shall be added to the Contract Price, and/or

(b) to terminate the Contract by giving 30 days' notice to the Engineer and the Purchaser, whether or not the Contractor has previously stopped work under paragraph (a) of this sub-clause.

Claims

Notification of claims

41.1 In every case in which circumstances arise which the Contractor considers entitle him, by virtue of the Conditions, to claim additional payment the following provisions shall take effect:

(a) within 30 days of the said circumstances arising the Contractor shall, if he intends to make any claim for additional payment, give to the Engineer notice of his intention to make a claim and shall state the reasons by virtue of which he considers that he is entitled thereto;

(b) as soon as is reasonably practicable after the date of the notice given by the Contractor of his intention to make a claim for additional payment, and not later than the expiry of the last Defects Liability Period, the Contractor shall submit to the Engineer (with copies for transmission to the Purchaser) full particulars of and the actual amount of his claim. The Contractor shall thereafter promptly submit such further particulars as the Engineer may reasonably require to assess the value of the claim.

Allowance for profit on claims

41.2 In any case where under the provisions of
sub-clause 5.7 (Unexpected site conditions),
sub-clause 11.8 (Breach of purchaser's general obligations),
sub-clause 14.5 (Revision of programme),
sub-clause 16.2 (Errors in drawings, etc. supplied by purchaser or engineer),
sub-clause 21.2 (special loads),
sub-clause 22.1 (Setting out),
sub-clause 25.2 (Additional cost caused by suspension),
sub-clause 25.5 (Resumption of work, delivery or erection),
sub-clause 25.6 (Effect of suspension on defects liability),
sub-clause 27.5 (Notice and confirmation of variations),
sub-clause 31.2 (Tests during defects liability period),
sub-clause 33.3 (Mitigation of consequences of delay),
sub-clause 36.8 (Contractor to search),
sub-clause 40.3 (Remedies on failure to certify or make payment) or
sub-clause 52.2 (Performance to continue during arbitration)

the Contractor is entitled to be paid or to have included in the Contract Price any extra or additional Cost the Contractor shall add to such Cost on account of profit the percentage thereof stated in the Appendix.

Purchaser's liability to pay claims

41.3 Notwithstanding anything contained in the Conditions the Purchaser shall not be liable to make payment in respect of any claim for an additional payment unless the Contractor has complied with the requirements of this clause.

Patent rights, etc.

Indemnity against infringement

42.1 The Contractor shall indemnify the Purchaser against all actions, claims, demands, costs, charges and expenses arising from or incurred by reason of any infringement or alleged infringement of patent, registered design, unregistered design right, copyright, trade mark or trade name protected in the country where the Plant is to be manufactured or where the Works are to be erected by the use or possession of any Plant supplied by the Contractor, but such indemnity shall not cover any use of the Works otherwise than for the purpose indicated by or reasonably to be inferred from the Specification or any infringement which is due to the use of any Plant in association or combination with any other plant not supplied by the Contractor.

Conduct of proceedings

42.2 In the event of any claim being made or action brought against the Purchaser arising out of the matters referred to in this clause, the Contractor shall be promptly notified thereof and may at his own expense conduct all negotiations for the settlement of the same and any litigation that may arise therefrom. The Purchaser shall not, unless and until the Contractor shall have failed to take over the conduct of the negotiations or litigation, make any admission which might be prejudicial thereto. The conduct by the Contractor of such negotiations or litigation shall be conditional upon the Contractor having first given to the Purchaser such reasonable security as shall from time to time be required by the Purchaser to cover the amount ascertained or agreed or estimated, as the case may be, of any compensation, damages, costs, charges and expenses for which the Purchaser may become liable. The Purchaser shall, at the request of the Contractor, afford all available assistance for the purpose of contesting any such claim or action, and shall be repaid all reasonable expenses incurred in so doing.

Purchaser's indemnity against infringement

42.3 The Purchaser on his part warrants that any design or instructions furnished or given by him or by the Engineer on his behalf shall not be such as will cause the Contractor to infringe any patent, registered design, unregistered design right, copyright, trade mark or trade name in the performance of the Contract and shall indemnify the Contractor in the same terms as the Contractor indemnifies the Purchaser under sub-clause 42.1 (Indemnity against infringement). The provisions of sub-clause 42.2. (Conduct of proceedings) shall apply with the necessary changes of detail being made.

Infringement preventing performance

42.4 If the Contractor shall be prevented from executing the Works, or the Purchaser is prevented from using the Works, in consequence of any infringement of patent, registered design, unregistered design right, copyright, trade mark or trade name and the party indemnifying the other in accordance with sub-clause 42.1 (Indemnity against infringement) or sub-clause 42.3 (Purchaser's indemnity against infringement) is unable within 90 days after notice thereof from the other party to procure the removal at his own expense of the cause of prevention then:

(a) in the case of an infringement which is the subject of the Contractor's indemnity to the Purchaser under sub-clause 42.1 (Indemnity against infringement) the Purchaser may treat such prevention as a default by the Contractor and exercise the powers and remedies available to him under clause 49 (Contractor's default), and

(b) in the case of an infringement which is the subject of the Purchaser's indemnity under sub-clause 42.3 (Purchaser's indemnity against infringement) the Contractor may treat such prevention as a default by the Purchaser and exercise the powers and remedies available to the Contractor under clause 51 (Purchaser's default).

Copyright © 2000 Institution of Electrical Engineers

Accidents and damage

Care of the works

43.1 The Contractor shall be responsible for the care of the Works or any Section thereof until the date of taking-over as stated in the Taking-Over Certificate applicable thereto. The Contractor shall also be responsible for the care of any outstanding work which he has undertaken to carry out during the Defects Liability Period until such outstanding work is complete. In the event of termination of the Contract in accordance with the Conditions, responsibility for the care of the Works shall pass to the Purchaser upon expiry of the notice of termination, whether given by the Purchaser or by the Contractor.

Making good loss or damage to the works

43.2 In the event that any part of the Works shall suffer loss or damage whilst the Contractor has responsibility for the care thereof, the same shall be made good by the Contractor at his own expense except to the extent that such loss or damage shall be caused by the Purchaser's Risks. The Contractor shall also at his own expense make good any loss or damage to the Works occasioned by him in the course of operations carried out by him for the purpose of completing any outstanding work or of complying with his obligations under clause 36 (Defects liability).

Damage to works caused by purchaser's risks

43.3 In the event that any part of the Works shall suffer loss or damage whilst the Contractor has responsibility for the care thereof which is caused by any of the Purchaser's Risks the same shall, if required by the Purchaser within six months after the happening of the event giving rise to loss or damage, be made good by the Contractor. Such making good shall be at the expense of the Purchaser at a price to be agreed between the Contractor and the Purchaser. In default of agreement such sum as is in all the circumstances reasonable shall be determined by arbitration under clause 52 (Disputes and arbitration). The price or sum so agreed or determined shall be added to the Contract Price.

Injury to persons or damage to property whilst contractor has responsibility for care of the works

43.4 Except as hereinafter mentioned the Contractor shall be liable for and shall indemnify the Purchaser against all claims in respect of personal injury or death or in respect of loss of or damage to any property (other than property forming part of the Works not yet taken over) which arises out of or in consequence of the execution of the Works whilst the Contractor has responsibility for the care thereof and against all demands, costs, charges and expenses arising in connection therewith. The Contractor shall not be liable under this sub-clause for, and the Purchaser shall indemnify him from and against, any claims in relation to death or personal injury or loss of or damage to property to the extent that the same is caused by any of the Purchaser's Risks and in the case of damage to property to the further extent that the damage is an inevitable consequence of the execution of the Works.

Injury to persons or damage to property after responsibility for care of the works passes to purchaser

43.5 If there shall occur any death or injury to any person or loss of or damage to any property (other than the Works) after the responsibility for the care of the Works shall have passed to the Purchaser the Contractor shall be liable for and shall indemnify the Purchaser against all such claims and all actions, demands, costs, charges and expenses arising in connection therewith to the extent that such death or personal injury or loss of or damage to property was caused by the negligence or breach of statutory duty of the Contractor, his Sub-Contractors, servants or agents or by defective design [other than a design for which the Contractor has disclaimed responsibility in accordance with sub-clause 13.3 (Contractor's design)], materials or workmanship but not otherwise. The Contractor's liability for any loss or damage to the Works shall be limited to the fulfilment of his obligations in relation thereto under clause 36 (Defects liability).

Accidents or injury to workmen

43.6 The Contractor shall indemnify the Purchaser against all actions, suits, claims, demands, costs, charges and expenses arising in connection with the death of or injury to any person employed by the Contractor or his Sub-Contractors for the purposes of the Works. This indemnity shall not apply to the extent that any death or injury results from any act or default of the Purchaser, his servants, agents or other contractors for whom he is responsible. The Purchaser shall indemnify the Contractor against all claims, damages, costs, charges and expenses to such extent.

Copyright © 2000 Institution of Electrical Engineers

Claims in respect of injury to persons or damage to property

43.7 In the event of any claim being made against the Purchaser arising out of the matters referred to in this clause and in respect of which it appears that the Contractor may be liable under this clause the Contractor shall be promptly notified thereof and may at his own expense conduct all negotiations for the settlement of the same and any litigation that may arise in relation thereto. The Purchaser shall not, unless and until the Contractor shall have failed to take over the conduct of the negotiations or litigation, make any admission which might be prejudicial thereto. The conduct by the Contractor of such negotiations or litigation shall be conditional upon the Contractor having first given to the Purchaser such reasonable security as shall from time to time be required by him to cover the amount ascertained or agreed or estimated, as the case may be, of any compensation, damages, costs, charges and expenses for which the Purchaser may become liable. The Purchaser shall, at the request of the Contractor, afford all available assistance for any such purpose and shall be repaid all expenses reasonably incurred in so doing.

Limitations of liability

Mitigation of loss

44.1 In all cases the party establishing or alleging a breach of contract or a right to be indemnified in accordance with the Contract shall be under a duty to take all necessary measures to mitigate the loss which has occurred provided that he can do so without unreasonable inconvenience or cost.

Indirect or consequential damage

44.2 Except as expressly provided in sub-clauses 34.1 (Delay in completion) and 35.8 (Consequences of failure to pass performance tests) for the payment or deduction of liquidated damages for delay or failure to achieve performance and except for those provisions of the Conditions whereby under sub-clause 41.2 (Allowance for profit on claims) the Contractor is expressly stated to be entitled to receive profit, neither the Contractor nor the Purchaser shall be liable to the other by way of indemnity or by reason of any breach of the Contract or of statutory duty or by reason of tort (including but not limited to negligence) for any loss of profit, loss of use, loss of production, loss of contracts or for any financial or economic loss or for any indirect or consequential damage whatsoever that may be suffered by the other.

Limitation of contractor's liability

44.3 In no circumstances whatsoever shall the liability of the Contractor to the Purchaser under the Conditions for any one act or default exceed the sum stated in the Appendix or if no sum is so stated, the Contract Price. The Contractor shall have no liability to the Purchaser for or in respect or in consequence of any loss of or damage to the Purchaser's property which shall occur after the expiration of the Defects Liability Period except as stated in sub-clause 36.10 (Latent defects).

Exclusive remedies

44.4 The Purchaser and the Contractor intend that their respective rights, obligations and liabilities as provided for in the Conditions shall be exhaustive of the rights, obligations and liabilities of each of them to the other arising out of, under or in connection with the Contract or the Works, whether such rights, obligations and liabilities arise in respect or in consequence of a breach of contract or of statutory duty or a tortious or negligent act or omission which gives rise to a remedy at common law. Accordingly, except as expressly provided for in the Conditions, neither party shall be obligated or liable to the other in respect of any damages or losses suffered by that other which arise out of, under or in connection with the Contract or the Works, whether by reason or in consequence of any breach of contract or of statutory duty or tortious or negligent act or omission.

Purchaser's risks

Purchaser's risks

45.1 The 'Purchaser's Risks' are:

– fault, error, defect or omission in designs furnished or specified by the Purchaser or the Engineer responsibility for which has been disclaimed

- by the Contractor in the manner provided for by sub-clause 13.3 (Contractor's design);

- the use or occupation of the Site by the Works, or for the purposes of the Contract; interference, whether temporary or permanent with any right of way, light, air, or water or with any easement, wayleave or right of a similar nature which is the inevitable result of the construction of the Works in accordance with the Contract;

- damage (other than that resulting from the Contractor's method of construction) which is the inevitable result of the construction of the Works in accordance with the Contract;

- use of the Works or any part thereof by the Purchaser;

- the act, neglect or omission or breach of contract or of statutory duty of the Engineer or the Purchaser, his agents, servants or other contractors for whom the Purchaser is responsible;

- Force Majeure except to the extent insured under the insurance policies to be effected by the Contractor in accordance with clause 47 (Insurance).

Force majeure

Force majeure 46.1 'Force Majeure' means:

- war, hostilities (whether war be declared or not), invasion, act of foreign enemies;

- ionising radiations, or contamination by radio-activity from any nuclear fuel, or from any nuclear waste from the combustion of nuclear fuel, radio-active toxic explosive, or other hazardous properties of any explosive nuclear assembly or nuclear component thereof;

- pressure waves caused by aircraft or other aerial devices travelling at sonic or supersonic speeds;

- rebellion, revolution, insurrection, military or usurped power or civil war;

- riot, civil commotion or disorder;

- any circumstances beyond the reasonable control of either of the parties.

Notice of force majeure 46.2 If either party is prevented or delayed from or in performing any of his obligations under the Contract by Force Majeure, then he may notify the other party of the circumstances constituting the Force Majeure and of the obligations performance of which is thereby delayed or prevented, and the party giving the notice shall thereupon be excused the performance or punctual performance, as the case may be, of such obligations for so long as the circumstances of prevention or delay may continue.

Termination for force majeure 46.3 Notwithstanding that the Contractor may have been granted under sub-clause 33.1 (Extension of time for completion) an extension of the Time for Completion of the Works, if by virtue of sub-clause 46.2 (Notice of force majeure) either party shall be excused the performance of any obligation for a continuous period of 120 days, then either party may at any time thereafter, and provided that such performance or punctual performance is still excused, by notice to the other terminate the Contract.

Payment on termination for force majeure 46.4 If the Contract is terminated under sub-clause 46.3 (Termination for force majeure) the Engineer shall certify, and the Purchaser shall pay to the Contractor in so far as the same shall not have already been included in certificates of payment paid by the Purchaser or be the subject of an advance payment, the Contract Value of the Works executed prior to the date of termination.

The Contractor shall also be entitled to have included in a certificate of payment and to be paid:

(a) the Cost of materials or goods reasonably ordered for the Works or for use in connection with the Works which have been delivered to the Contractor or of which the Contractor is legally liable to accept delivery. Such materials or goods shall become the property of the Purchaser when paid for by the Purchaser. The Purchaser shall be entitled to withhold payment in respect thereof until such materials or goods have been delivered to, or to the order of, the Purchaser;

(b) the amount of any other expenditure which in the circumstances was reasonably incurred by the Contractor in the expectation of completing the whole of the Works;

(c) the reasonable Cost of removal of Contractor's Equipment and the return thereof to the Contractor's works in his country or to any other destination at no greater Cost;

(d) the reasonable Cost of repatriation of all the Contractor's staff and workmen employed at the Site on or in connection with the Works at the date of such termination.

Insurance

Insurance of works

47.1 The Contractor shall, in the joint names of the Contractor and the Purchaser, insure the Works and Contractor's Equipment and keep each part thereof insured for its full replacement value against all loss or damage from whatever cause arising, other than the Purchaser's Risks. Such insurance shall be effected from the date of the Letter of Acceptance, until 14 days after the date of issue of a Taking-Over Certificate in respect of the Works or any Section thereof, or if earlier, 14 days after the date when responsibility for the care of the Works passes to the Purchaser in accordance with the provisions of sub-clause 43.1 (Care of the works) upon expiry of notice of termination.

Extension of works insurance

47.2 The Contractor shall so far as reasonably possible extend the insurance under sub-clause 47.1 (Insurance of works) to cover damage which the Contractor is responsible for making good pursuant to clause 36 (Defects liability) or which occurs whilst the Contractor is on Site for the purpose of making good a defect or carrying out the Tests on Completion during the Defects Liability Period or supervising the carrying out of the Performance Tests or completing any outstanding work or which arises during the Defects Liability Period from a cause occurring prior to taking-over and for which the Contractor is liable under sub-clause 43.5 (Injury to persons or damage to property after responsibility for care of the works passes to purchaser).

Application of insurance monies

47.3 All monies received under any such policy shall be applied in or towards the replacement and repair of the part of the Works lost, damaged or destroyed but this provision shall not affect the Contractor's liabilities under the Contract.

Third party insurance

47.4 The Contractor shall, prior to the commencement of any work on the Site by the Contractor pursuant to the Contract, insure in an amount not being less than the amount stated in the Special Conditions against his liability for damage or death or personal injury occurring before all the Works have been taken over to any person (including any employee of the Purchaser) or to any property (other than property forming part of the Works) due to or arising out of the execution of the Works. The terms of the policy shall include a provision whereby, in the event of any claim being made against the Purchaser in respect of which the Contractor would be entitled to indemnity under the policy, the insurers will indemnify the Purchaser against such claims and any costs, charges and expenses in respect thereof.

Copyright © 2000 Institution of Electrical Engineers

Insurance against accident, etc. to workmen

47.5 The Contractor shall insure and shall maintain insurance against his liability under sub-clause 43.6 (Accidents or injury to workmen). The terms of any such policy shall also include the provision to indemnify the Purchaser mentioned in sub-clause 47.4 (Third party insurance) provided always that in respect of any persons employed by any Sub-Contractor, the Contractor's obligation under this sub-clause shall be satisfied if the Sub-Contractor shall have insured against the liability in respect of such persons in such manner that the Purchaser is indemnified under the policy, but the Contractor shall require such Sub-Contractor to produce to the Engineer when required the policy, the receipts for the premiums or other satisfactory evidence of insurance cover.

General insurance requirements

47.6 All insurances shall be effected with an insurer and in terms to be approved by the Purchaser (such approval not to be unreasonably withheld) and the Contractor shall from time to time, when so required by the Engineer, produce the policy and receipts for the premiums or other satisfactory evidence of insurance cover. The Contractor shall promptly notify the Purchaser of any alteration to the terms of the policy or in the amounts for which insurance is provided.

Exclusions from insurance cover

47.7 The insurance policy may exclude cover for any of the following:

(a) the cost of making good or repairing any Plant which is defective or work which is not in accordance with the Contract;

(b) the Purchaser's Risks;

(c) indirect or consequential loss or damage including any deductions from the Contract Price for delay;

(d) fair wear and tear; shortages and pilferages;

(e) risks related to mechanically propelled vehicles for which third party or other insurance is required by law.

Remedy on failure to insure

48.1 If the Contractor shall fail to effect and keep in force the insurances referred to in the Conditions the Purchaser may effect and keep in force any such insurance and pay such premiums as may be necessary for that purpose and from time to time deduct the amount so paid by the Purchaser from any monies due or which may become due to the Contractor under the Contract or recover the same as a debt from the Contractor.

Joint insurances

48.2 Wherever insurance is arranged under the Conditions in the joint names of the parties, or on terms containing provisions for indemnity to principals, the party effecting such insurance shall procure that the subrogation rights of the insurers against the other party are waived and that such policy shall permit either:

(a) the co-insured, or

(b) the other party to the Contract

to be joined to and be a party to any negotiations, litigation or arbitration upon the terms of the policy or any claim thereunder.

Contractor's default

Contractor's default

49.1 If the Contractor shall assign the Contract, or sub-let the whole of the Works without the consent of the Purchaser, or if the Engineer has rejected the Works or a Section under sub-clauses 28.5 (Consequences of failure to pass tests on completion) or 35.8 (Consequences of failure to pass performance tests) or shall certify that the Contractor:

(a) has abandoned the Contract, or

Copyright © 2000 Institution of Electrical Engineers

(b) has without reasonable excuse suspended the progress of the Works for 30 days after receiving from the Engineer written notice to proceed, or

(c) despite previous warnings in writing from the Engineer is not executing the Works in accordance with the Contract, or is failing to proceed with the Works with due diligence or is neglecting to carry out his obligations under the Contract so as to affect adversely the carrying out of the Works,

then the Purchaser may give 21 days' notice to the Contractor of his intention to proceed in accordance with the provisions of this clause. Upon the expiry of such notice the Purchaser may without prejudice to any other remedy under the Contract forthwith terminate the Contract and enter the Site and expel the Contractor therefrom but without thereby releasing the Contractor from any of his obligations or liabilities which have accrued under the Contract and without affecting the rights and powers conferred by the Contract on the Purchaser or the Engineer. Upon such termination the Purchaser may himself complete the Works or may employ any other contractor so to do, and the Purchaser shall have the free use of any Contractor's Equipment for the time being on the Site.

Valuation at date of termination

49.2 As soon as practicable after the Purchaser has terminated the Contract the Engineer shall, by or after reference to the parties and after making such enquiries as he thinks fit, value the part of the Works executed prior to the date of termination and all sums then due to the Contractor as at the date of termination in accordance with the principles of clause 39 (Certificates and payment) and certify the amount thereof. The amount so certified is herein called 'the Termination Value'.

Payment after termination

49.3 The Purchaser shall not be liable to make any further payments to the Contractor until the Cost of execution and all other expenses incurred by the Purchaser in completing the Works have been ascertained and the amount payable certified by the Engineer (herein called 'the Cost of Completion'). If the Cost of Completion when added to the total amounts already paid to the Contractor as at the date of termination exceeds the total amount which the Engineer certifies would have been payable to the Contractor for the execution of the Works, the Engineer shall certify such excess and the Contractor shall upon demand pay to the Purchaser the amount of such excess. Any such excess shall be deemed a debt due by the Contractor to the Purchaser and shall be recoverable accordingly. If there is no such excess the Contractor shall be entitled to be paid the difference (if any) between the Termination Value and the total of all payments received by the Contractor as at the date of termination.

Bankruptcy and insolvency

50.1 If the Contractor becomes bankrupt or insolvent, or has a receiving order made against him, or compounds with his creditors, or, being a corporation, commences to be wound up (not being a members' voluntary winding up for the purposes of amalgamation or reconstruction) or has an administration order made against him or carries on his business under an administrator, a receiver, a manager or liquidator for the benefit of his creditors or any of them, the Purchaser shall be entitled:

(a) to terminate the Contract forthwith by notice to the Contractor or to the administrator, receiver, manager or liquidator or to any person in whom the Contract may become vested, in which event the provisions of clause 49 (Contractor's default) shall apply, or

(b) to give such administrator, receiver, manager or liquidator or other person the option of carrying out the Contract subject to his providing a guarantee for the due and faithful performance of the Contract up to an amount to be agreed.

Purchaser's default

Notice of termination due to purchaser's default

51.1 In the event of the Purchaser:

(a) failing to pay to the Contractor the amount due under any certificate of the Engineer within 30 days after the date of its issue subject to any deduction that the Purchaser is entitled to make under the Contract, or

(b) interfering with or obstructing the issue of any certificate of the Engineer, or

(c) becoming bankrupt or (being a corporation) going into liquidation other than for the purpose of a scheme of reconstruction or amalgamation, or carrying on its business under an administrator, receiver, manager or liquidator for the benefit of its creditors or any of them, or

(d) appointing a person to act with or in replacement of the Engineer against the reasonable objections of the Contractor,

the Contractor shall be entitled without prejudice to any other rights or remedies under the Contract [and in respect of paragraph (a) above in addition to the provisions of sub-clause 40.3 (Remedies on failure to certify or make payment)] to terminate the Contract by giving 14 days' notice to the Purchaser with a copy to the Engineer.

Removal of contractor's equipment

51.2 Upon the giving of notice under sub-clause 51.1 (Notice of termination due to purchaser's default) the Contractor shall with all reasonable despatch remove from the Site all Contractor's Equipment.

Payment on termination due to purchaser's default

51.3 In the event of termination under sub-clause 51.1 (Notice of termination due to purchaser's default) the Engineer shall act as provided in sub-clause 49.2 (Valuation at date of termination) and certify the Termination Value of the Works as at the date of termination. The Engineer shall, on the application of the Contractor accompanied by supporting details, also certify the amount of any expenditure reasonably incurred by the Contractor in the expectation of the performance of, or in consequence of the termination of, the Contract to the extent that the same has not been included in the Termination Value. The Engineer shall also certify in respect of the Contractor's loss of anticipated profit on the Contract the percentage referred to in sub-clause 41.2 (Allowance for profit on claims) on the difference between the total of the Termination Value plus the expenditure before referred to and the Contract Price but in no case shall the total amounts so certified exceed the Contract Price. Thereafter the Engineer shall issue a certificate of payment for the amount by which the said Termination Value, expenditure and allowance for profit exceeds the total of sums previously paid to the Contractor and such certificate of payment shall be paid by the Purchaser within 30 days after the date of issue.

Disputes and arbitration

Notice of arbitration

52.1 If at any time any question, dispute or difference shall arise between the Purchaser and the Contractor in relation to the Contract which cannot be settled amicably, either party shall as soon as is reasonably practicable give to the other notice of the existence of such question, dispute or difference specifying its nature and the point at issue, and the same shall be referred to the arbitration of a person to be agreed upon. Failing agreement upon such person within 30 days after the date of such notice, the arbitration shall be conducted by some person appointed on the application of either party by the President of the institution named in the Appendix (or by his deputy appointed by such President for the purpose). A question, dispute or difference relating to a decision, instruction or order of the Engineer shall not be referred to arbitration except in accordance with sub-clause 2.6 (Disputing engineer's decisions, instructions and orders).

Performance to continue during arbitration

52.2 Performance of the Contract shall continue during arbitration proceedings unless the Engineer shall order the suspension thereof. If such suspension be ordered the additional Cost to the Contractor occasioned by such suspension shall be added to the Contract Price. No payment due or payable by the Purchaser shall be withheld on account of a pending reference to arbitration.

Arbitrator's powers

52.3 The arbitrator shall have power:

(a) to open up, review and revise any certificate or valuation of the Engineer or any decision, opinion, instruction or order of the Engineer referred to arbitration under sub-clause 2.6 (Disputing engineer's decisions, instructions and orders);

(b) to order on a provisional basis (subject to the arbitrator's final award) any relief which he would have power to grant in a final award including but not limited to the making of a provisional order for the payment of money as between the parties or an order to make an interim payment on account of the costs of the arbitration.

Joinder

52.4 Where a dispute has been referred to arbitration under this clause and the Contractor is in a related dispute with a Sub-Contractor which is substantially the same as the matter referred to arbitration hereunder, the parties consent to the joinder of the Sub-Contractor as a party to the arbitration and to the reference of such related dispute to the arbitrator appointed hereunder and further agree that the arbitrator shall have power to order the consolidation of such arbitration proceedings and/or to order the holding of concurrent hearings.

Arbitration rules

52.5 The arbitration shall be conducted in accordance with the arbitration rules specified in the Special Conditions or, if no rules be so specified, in accordance with the provisions of the Arbitration Act 1996.

Sub-contractors, etc.

Sub-contractors, servants and agents

53.1 It is expressly agreed that no servant or agent of the Contractor nor any Sub-Contractor shall in any circumstances whatsoever (with the exception of liability for death or personal injury caused by wilful or negligent acts or omissions) be under any obligation, responsibility or liability to the Purchaser for or in respect of any loss, damage or injury of whatsoever kind and howsoever arising. Without prejudice to the generality of the foregoing every limitation and exclusion of liability of the Contractor contained in the Conditions shall also extend to protect every such servant, agent or Sub-Contractor. For the purposes of this clause the Contractor is or shall be deemed to be acting as agent or trustee on behalf of and for the benefit of all persons who are or who may from time to time become servants, agents or Sub-Contractors as aforesaid and to such extent all such persons shall be or be deemed to be parties to the Contract.

Applicable law

Applicable law

54.1 Unless otherwise stated in the Special Conditions, the Contract shall in all respects be governed by and interpreted in accordance with the laws of England and English law shall govern the procedure of any arbitration under clause 52 (Disputes and arbitration).

GENERAL CONDITIONS

APPENDIX

Prime cost items SUB-CLAUSE 5.5

Percentage to be added .... %

Delay in completion SUB-CLAUSE 34.1

Percentage of Contract Value to be paid or deducted for each week of delay .... %

Maximum percentage of Contract Value which payments or deductions shall not exceed .... %

(Insert the percentage of the Contract Value for each week of delay and the maximum percentage of Contract Value to be paid or deducted for each Section of the Works where more than one Section has been defined for the purpose of taking-over.)

Prolonged delay SUB-CLAUSE 34.2

Maximum loss recoverable by the Purchaser £ ....

Allowance for profit on claims SUB-CLAUSE 41.2

Percentage to be added .... %

Limitation of contractor's liability SUB-CLAUSE 44.3

Limit of liability £ ....

Person to appoint arbitrator SUB-CLAUSE 52.1

The President of the Institution of Mechanical Engineers

The President of the Institution of Electrical Engineers

(delete as appropriate)

SPECIAL CONDITIONS
Aide-mémoire to their preparation

Numbers refer to the general conditions' sub-clause numbers.

Purchaser	**1.**1.a	The Purchaser is ....
Engineer	**1.**1.d	The Engineer is ....

Time for completion **1.**1.m [Insert Time for Completion of the Works and, where appropriate, Time for Completion of each Section of the Works.]

Engineer's duties **2.**1 The Engineer is required to obtain the Purchaser's prior specific approval before exercising the following duties:

(List as appropriate.)

Changes in costs–labour, materials and transport

6.2 Either (a) sub-clause 6.2 of the general conditions shall not apply

or (b) variations in the cost of labour, material and transport and adjustments to the Contract Price occasioned thereby shall be calculated in accordance with the following formula(e):

(Set out formula(e) applicable.)

Performance bond or guarantee **8.**1

Bond Amount	[..........................]	(% of Contract Price)
Currency of bond	[............…................]	(eg. sterling)
Period of validity	[................….............]	(Expiry Date)
Procedure for forfeiture	[............…................]	(claims on bond)
Arrangements for release	[..................….....]	(automatic/on expiry/other)

Notices
10.1 Purchaser's address: ....
Engineer's address: ....
10.2 Contractor's address: ....

Purchaser's lifting equipment **11.**5 The following lifting equipment belonging to the Purchaser may be used by the Contractor in connection with the Works:

[Here set out list of equipment and terms and conditions of use.]

Utilities and power **11.**6 Utilities available for use by the Contractor on Site are as follows:

[Here list utilities available. Specify points from which each is available and charges to be levied against the Contractor for use in accordance with the general conditions, sub-clause 18.2 (Site services).]

Power, etc. for tests on site **11.**7 Specify items which will not be provided and/or the use of which will be charged to the Contractor.

Form of programme

14.2 The Programme shall be in the following form:

[Here set out form required. It must include a clear indication of the Time for Completion of the Works and, where appropriate, of the Time for Completion of each Section.]

Performance tests

35.1 Specify tests required in detail and performance to be obtained and the acceptance limits.

35.6 Time for completion of Performance Tests (eg. [......................] months after taking-over).

35.8 Liquidated damages for failure to pass Performance Tests:-

(eg. £ [.......................] per percentage point loss of guaranteed efficiency).

(Separate liquidated damages should be identified for each Section of the Works for which Performance Tests have been specified.)

Defects liability period

36.1 The Defects Liability Period shall be [...................] months after taking-over.

Certificates and payment

39.2 Form of application:

(Set out evidence of value of payment of freight and insurance (eg. copy invoices) and documentation required to accompany applications for payment including where necessary evidence of Contractor's right to transfer ownership.)

Payment

40.1 (Set out terms of payment including entitlement to advance or progress payments. The following suggested clauses (adapted as required) may be used.)

Progress certificates of payment

1. The Contractor may make application to the Engineer for first, second and third progress certificates of payment when the value of Plant manufactured and work executed before delivery to Site amounts to not less than [25]%, [50]% and [75]% respectively of the Contract Price and the Contractor has furnished to the Engineer reasonable evidence of such value and of the Contractor's right to transfer the property therein to the Purchaser, and that he has suitably and sufficiently marked such part of the Plant as the Engineer may reasonably require as the property of the Purchaser. Not more than three applications for such certificates may be made except in accordance with sub-clause 3 of this clause, and the Contractor being otherwise entitled to make application for a second or third such certificate shall not be debarred from making such application on account of his not having applied for a first or second progress certificate of payment as the case may be.

2. Every progress certificate of payment shall certify that the total value of the Plant manufactured and work executed up to the date of the certificate amounts to not less than [25]%, [50]% or [75]% of the Contract Price as the case may be, and for the purposes of sub-clause 39.4 (Value included in certificates of payment) the total value certified in a progress certificate of payment shall be deemed to be [25]%, [50]% or [75]% of the Contract Price as the case may be.

3. In the event of the Works being divided by the Contract into two or more Sections the provisions of sub-clauses 1 and 2 of this clause shall have effect as if each Section formed the subject of a separate Contract and as if the expression 'the Contract Price' meant the part of the Contract Price properly attributable to the Section.

4. The provisions of sub-clauses 39.3 (Issue of payment certificate), 39.5 (Adjustments to certificates) and 39.8 (Effect of certificates of payment) shall have effect in relation to progress certificates of payment in all respects as if such certificates were interim certificates of payment.

Terms of payment

1. The Purchaser shall pay to the Contractor in the following manner the Contract Price adjusted to give effect to such additions thereto and such deductions therefrom as are provided for in these Conditions:

(a) [10]% of the Contract Price, as an advance payment, within 30 days after the Contractor has furnished to the Purchaser an irrevocable letter of guarantee, from a guarantor or surety acceptable to the Purchaser, with a value equivalent to the advance payment. The letter of guarantee shall provide for its value to reduce by an amount equal to [10]% of the sum certified in each interim certificate of payment.

(b) Within 30 days after presentation of each interim certificate of payment a sum equal to [85]% of the sum certified therein in respect of Plant shipped or delivered to Site (as the case may be) and of freight and insurance paid in respect of Plant shipped,

and

[85]% of the sum certified therein in respect of work done on Site.

(c) [2☐]% of the Contract Price adjusted as aforesaid within 30 days after presentation of the Taking-Over Certificate.

(d) The balance of the Contract Price adjusted as aforesaid within 30 days after presentation of the final certificate of payment. If the Contractor shall have furnished to the Purchaser a guarantee acceptable to the Purchaser for the payment on demand of such balance, he shall be entitled to payment thereof with or at any time after the payment provided for by paragraph (c) hereof.

2. If any Section of the Works shall be taken over separately under clause 29 (Taking-over) the payments herein provided for on or after taking-over shall be made in respect of the Section taken over and reference to the Contract Price shall mean such part of the Contract Price as shall, in the absence of agreement, be apportioned thereto by the Engineer.

3. If at any time at which payment would fall to be made under paragraph (b) or paragraph (c) of sub-clause 1 of this clause there shall be any defect in any portion of the Works in respect of which such payment is proposed, the Purchaser may retain the whole of such payment provided that, in the event of the said defect being of a minor character and not such as to affect the use of the Works or the said portion thereof for the purpose intended without serious risk, the Purchaser shall not retain a greater sum than represents the Cost of making good the said minor defect. Any sum retained by the Purchaser pursuant to the provisions of this sub-clause shall be paid to the Contractor upon the said defect being made good.

Delayed payment

40.2 (Where the rate specified in sub-clause 40.2 is not to apply.)

The rate of interest on overdue payment shall be [..................]% per annum.

Third party insurance

47.4 The Contractor shall effect third party insurance for a minimum amount of £ [.....................] for any one claim or series of claims arising from the same incident.

Copyright © 2000 Institution of Electrical Engineers

Disputes and arbitration

Arbitration rules

52.5 The arbitration shall be conducted in accordance with the [arbitration rules of [1] insert body, eg. International Chamber of Commerce (ICC), United Nations Commission on International Trade Law (UNCITRAL), London Court of International Arbitration (LCIA), Chartered Institute of Arbitrators (CIArb) or other] Construction Industry Model Arbitration Rules (CIMAR)

[1] choose as appropriate

Scots law

If sub-clause 54.1 (Applicable law) below is completed so as to make the Contract subject to Scots law:

(i) sub-clause 52.1 (Notice of arbitration) should be amended by adding the following sentence before the last sentence of the sub-clause:

Where the Special Conditions provide that the Contract is subject to Scots law, such person shall be a single arbiter and the decision of the arbiter shall be final, it being agreed that neither party shall invoke the provisions of Section 3 of the Administration of Justice (Scotland) Act 1972.

(ii) sub-clause 52.3 (Arbitrator's powers) should be amended by changing references to "arbitrator" to "arbiter" and inserting the following wording as a new sub-paragraph (b) and by the re-lettering of the existing sub-paragraph (b) as (c):

(b) to order that damages, costs and/or expenses are due and payable as between the parties and to assess the amount of such damages, costs and/or expenses:

The re-lettered sub-paragraph (c) should be amended by the insertion of the following wording after the words " ... order for the payment of money ... ": " ... by way of damages or otherwise ... "

(iii) the heading of sub-clause 52.4 (Joinder) should be "Third parties" in place of "Joinder" and reference to "joinder of the Sub-Contractor as a party" should, instead, be to the "Sub-Contractor being brought in as a third party" to the arbitration. References to "arbitrator" should be changed to "arbiter".

(iv) the governing arbitration rules for the purposes of sub-clause 52.5 (Arbitration rules) should be chosen from the following selection:

(1) The Arbitration Rules of the International Chamber of Commerce as convened in Edinburgh.

(2) The Arbitration Rules of UNCITRAL as applicable in Scotland by virtue of the Law Reform Miscellaneous Provisions (Scotland) Act 1990, Section 66 and Schedule 7.

(3) The Arbitration Rules of the Law Society of Scotland.

Add at the end of this sub-clause 52.5: " ... except to the extent otherwise provided or specified in clause 52."

(v) an additional sub-clause 52.6 (Consent to registration) should be added:

Consent to registration

52.6 The parties consent to the registration in the Books of Council and Session for preservation and execution of any decree arbitral or award issued pursuant to arbitration proceedings under clause 52.

Applicable law If it is desired that the laws of a country other than England should govern the Contract, appropriate advice should be taken as to the legal effect of so doing. In such circumstances the following sub-clause duly completed may be a suitable alternative:

54.1 The Contract shall in all respects be governed by and interpreted in accordance with the laws of [..........................] and [................................] law shall govern the procedures of any arbitration under clause 52 (Disputes and arbitration).

Additional special conditions as appropriate

Labour Permits, Housing and standards of accommodation, Safety and health, etc.

Taxation Payment or exemption of contractor and his employees from local or other taxes.

Miscellaneous Bribery and corruption, Photographs and publicity of the works.

Foreign currency Arrangement for payments in currencies other than sterling, rates of exchange for conversion, etc.

ADDITIONAL SPECIAL CONDITIONS FOR USE IN CONTRACTS INVOLVING THE INCIDENTAL SUPPLY OF HARDWARE AND SOFTWARE

If the Works include the provision of computer hardware and software, for example, for a computerised control system, some or all of the following additional provisions may be considered appropriate.

1. DEFINITIONS

'Acceptance' means in relation to any part of the System the passing following commissioning of the Acceptance Tests relating thereto and 'accept' has a corresponding meaning.

'Acceptance Tests' means the tests of the System specified in the Schedule hereto or such other tests as the parties may agree pursuant to the Contract and which form part of the Tests on Completion applicable to the Works.

'Commissioning' means the installing of any part of the System and the preparation of such part for the Acceptance Tests and 'commission' has a corresponding meaning.

'Bespoke Software' means that part of the Software listed in the Schedule hereto to be developed by the Contractor under the Contract including all documentation in relation thereto to be provided by the Contractor under the Contract.

'Standard Software' means the Contractor's standard computer programs listed in the Schedule hereto including all documentation relating thereto to be provided by the Contractor under the Contract.

'Software' means the Bespoke Software and the Standard Software.

'Hardware' means the computer or microprocessor forming part of the Plant and all the equipment and operating systems as listed in the Schedule hereto.

'Software System Specification' means the specification (prepared by the Contractor and approved by the Engineer) of the design and means of implementation of each part of the Software.

'System' means that part of the Works which consists of the Hardware and the Software generally as described in the Statement of Requirements (as the same may be modified by the Software System Specification).

'Functional Specification' means the specification of the functions to be performed by the System prepared by the Contractor and approved by the Engineer.

'Statement of Requirements' means the statement of requirements for the System prepared by the Engineer on behalf of the Purchaser and initialled by the parties.

'Trial Period' means in relation to the Software or any part thereof the period specified in the Statement of Requirements commencing when the Engineer is notified that such Software is available for use by the Purchaser for its intended purpose.

'the Premises' means that space or area within the Site of the Works within which the System is to be installed.

'Services' means the maintenance and operational support services (if any) to be provided by the Contractor in relation to the System, the Hardware and the Software brief particulars of which are given in the Schedule hereto.

2. WARRANTY AND PERFORMANCE

The Contractor warrants that the System and the Services shall fulfil the requirements and functions stated in the Statement of Requirements and the Functional Specification. In the event of conflict between the Statement of Requirements and the Functional Specification, the Statement of Requirements shall prevail.

3. PREPARATION OF PREMISES

If the Works do not include for the preparation of the Premises by the Contractor for the installation of the System the Contractor shall provide the Engineer with adequate information to enable the Purchaser to prepare the Premises for the System and to provide a suitable supply of electric current and any necessary mains services and all other required electrical and mechanical items (other than the Hardware and Software) and such environmental conditions as are necessary for the purposes of the System. The Purchaser shall at his own expense ensure that such preparation and provisions are made in accordance with the Programme sufficient to ensure that such work is complete before delivery of the Hardware.

4. ENGINEER'S ROLE

The Engineer's duties are contained in clause 2.1 (Engineer's duties) of the general conditions.

The Engineer shall be the single channel of communication between the Purchaser and the Contractor and no variation to the Software System Specification shall be binding upon the Purchaser unless approved in writing by the Engineer.

5. STEERING MEETINGS

The parties shall meet together with such frequency as may be desirable for the purposes, inter alia, of monitoring progress of that part of the Works which consists of the System, agreeing a delivery schedule for the Hardware, agreeing the Software System Specification, agreeing the Acceptance Tests therefor and the services and facilities (if any) to be provided by the Purchaser. Following approval in writing of the Software System Specification or any part thereof by the Engineer no amendment thereto shall be made by the Contractor unless specifically requested in writing so to do by the Engineer who shall signify his approval thereof in writing.

6. INSTALLATION OF HARDWARE

The Contractor shall, unless otherwise instructed by the Engineer, deliver the Hardware in accordance with the Programme or, if no time or date be stated in the Programme, within a reasonable time to enable the System to pass the Acceptance Tests within the Time for Completion of the Works. Following delivery of the Hardware the Contractor shall install the same and conduct the agreed tests and verification routines for the Hardware prior to the installation of the Software.

7. SOFTWARE AND SYSTEM ACCEPTANCE

Following successful passing of the tests and routines for the Hardware the Contractor shall provide, deliver, install and commission the Software. Each item representing part of the Standard Software and each part of the Bespoke Software shall be accepted when it has in conjunction with the Hardware successfully passed the Acceptance Tests specified for such part. The System shall be accepted when the Software has demonstrated its capacity to fulfil the requirements of the Functional Specification during the Trial Period. The general conditions shall apply to any such Acceptance Tests as if they were Tests on Completion. The Engineer shall issue a Taking-Over Certificate for the System under clause 29 (Taking-over) of the general conditions when all the Acceptance Tests have been successfully passed. If the System fails to pass the Acceptance Tests the Engineer may in accordance with sub-clause 28.5 (Consequences of failure to pass tests on completion) of the general conditions reject the System.

8. TITLE TO STANDARD SOFTWARE

Title to and copyright in the Standard Software shall at all times remain vested in the Contractor. The Contractor shall grant to the Purchaser the non-exclusive right to use the Standard Software in the System for the purposes of his business. The Purchaser shall be entitled to make such copies of the Standard Software as he may reasonably require for his own internal use without the payment of any royalty to the Contractor and will ensure that all such copies acknowledge the Contractor's copyright therein and in the product thereof. The Purchaser shall take all reasonable measures to safeguard the Contractor's rights in the Standard Software.

9. TITLE TO BESPOKE SOFTWARE

Title to and copyright in the Bespoke Software shall vest in the Purchaser. The Purchaser shall at the request of the Contractor grant to the Contractor a non-exclusive worldwide licence to market the Bespoke Software (and to sub-licence third parties to use the same) upon such terms as may be agreed, or in the absence of agreement, upon such terms as may be fixed by arbitration under clause 52 (Disputes and arbitration) of the general conditions as being in all the circumstances reasonable.

10. CONFIDENTIALITY

(a) The Contractor shall keep confidential any information obtained pursuant to this Contract from the Purchaser or the Engineer and shall not divulge the same to any third party without the consent in writing of the Purchaser. The provisions of this clause shall not apply to any information which shall come into the public domain otherwise than by breach of the Contract, to information in the possession of the Contractor before disclosure, to information obtained from a third party who is free to divulge the same or to any information which it is reasonably necessary for the Contractor to disclose for the purposes of sub-licences of the Bespoke Software under the licence referred to in clause 9 (Title to bespoke software) hereof. The Contractor shall only divulge confidential information to those of the Contractor's employees as are directly involved or engaged for the purposes of the Contract and who need to know the same and will ensure that such employees are aware of and comply with these obligations as to confidentiality. The Contractor shall ensure that any of his sub-contractors or suppliers are bound by the requirements of this clause.

(b) The Purchaser shall, and shall procure that his officers, servants and agents (including the Engineer) shall, at all times, whether before or after the issue of the final certificate of payment, keep the Software confidential except in so far as the same shall come into the public domain otherwise than through the default of the Purchaser or the

Engineer and that except as may be necessary for the purposes of and as permitted by the Contract and for the operation of the System will not use or copy (whether in written form or in magnetic or optical digital form) the same nor disclose the same to any third party otherwise than with the express written consent of the Contractor.

11. SOURCE CODES

The Contractor shall upon delivery of any part of the Bespoke Software:

(a) provide for the retention on the Premises of a copy of the source code thereof both in written form and in a magnetic or optical digital form, or

(b) deposit a copy of such source code with a third party approved by the Purchaser upon terms that upon the Contractor going into liquidation or having a receiver appointed, or ceasing to provide service in respect thereof, the same shall on the written request of the Purchaser be released forthwith by such third party to the Purchaser for use by the Purchaser solely for the purposes of maintaining the System in connection with the Purchaser's business.

12. SERVICES

Where so specified in the Contract, following Acceptance of the System the Contractor shall offer to provide services of maintenance and advisory and consultancy services for the Hardware and Software including the provision of an updating service for the Standard Software as specified in the Statement of Requirements.

13. TRAINING

The Contractor shall provide instruction in the use of the Hardware and Software and the System for the Purchaser's personnel in accordance with the details given in the Schedule hereto.

14. MANUALS AND USER DOCUMENTATION

The Contractor shall provide as part of the Software all operation manuals, clerical control manuals, fault-finding procedures and routines and full operating and user manual documentation necessary for the use of the System in such numbers as the parties may agree. Notwithstanding the passing of any Acceptance Tests the System shall not be deemed to have been accepted nor shall the Engineer issue a Taking-Over Certificate in respect thereof until all such documentation shall have been provided to the Purchaser.

15. DEFECTS LIABILITY

(a) For a period of twelve months after Acceptance of the System the Contractor shall be responsible for making good with all possible speed at his own expense any defect in or damage to any portion of the Software which results in the failure of the System to fulfil the Statement of Requirements and Functional Specification and which arises from defective materials, including Software, workmanship or design (other than a design furnished or specified by the Purchaser and/or the Engineer and for which the Contractor has disclaimed responsibility in writing in accordance with sub-clause 13.3 (Contractor's design) of the general conditions).

(b) If any such damage or defect be not remedied within such time as may be agreed, the Purchaser may proceed to do the work at the Contractor's expense (such expense not to exceed the prices the Contractor would reasonably have charged for the work in question).

(c) Where a defect involves a fault inherent in the design of the System as specified in the Software System Specification, the Contractor shall at his own expense promptly carry out such redesign as may be necessary to prevent a recurrence of the defect and upon the completion of such redesign shall rectify the defect in the System. Any such redesign shall be accomplished in such a manner as to ensure that performance and operation of the System is not downgraded by virtue of such redesign from the standard shown on the passing of the Acceptance Tests.

(d) In all other respects the provisions of clause 36 (Defects liability) of the general conditions shall apply to any defect in or damage to the System which may occur during the Defects Liability Period.

16. PROVISIONS SURVIVING FINAL CERTIFICATES

The provisions of clauses 8 to 12 hereof shall where the context so requires continue in full force and effect notwithstanding the issue of a final certificate of payment in respect of the Works under clause 39 (Certificates and payment) of the general conditions and nothing in sub-clause 39.12 (Effect of final certificate of payment) of the general conditions shall affect the rights of the Purchaser or the obligations of the Contractor under the said clauses so continuing in full force and effect.

THE SCHEDULE

1. Acceptance Tests

2. Bespoke Software

3. Standard Software

4. Hardware

5. The Services

6. Training

ADDITIONAL SPECIAL CONDITIONS FOR USE IN CONTRACTS WHERE CERTIFICATION FOR PAYMENT AND PAYMENTS ARE TO BE DETERMINED IN WHOLE OR IN PART BY MEASUREMENT

Foreword

These additional Special Conditions are intended for use in contracts where certification for payment and payments are to be determined in whole or in part by measurement. The Special Conditions suggested include the revision of several sub-clauses in the MF/1 general conditions and the provision of additional clauses which are considered appropriate.

It is emphasised that the suggested additional Special Conditions are a model only and should be adapted and supplemented where necessary to reflect the nature of the Works and the problems likely to be encountered.

In particular, especially in relation to pipeline, powerline or cabling contracts, it may be necessary to expand the definition of the Site to reflect the fact that it will be over a lengthy route. The route, itself, will often be in a public area with limitations on access and requiring unique safety precautions. It may also be necessary to require the Contractor to record accurately the route and its depth or height in relation to a given datum. Whilst such requirements will normally be included in the Specification it may be appropriate to deal with them additionally in the Special Conditions.

Users of these additional Special Conditions are advised to take care when defining the units of work for which payment will be made and the method of measurement. Particular care should be taken to specify the full extent of each unit of work and to define clearly the boundaries between work units. Any general principles adopted in separating one unit of work from another should be clearly stated. Similarly, it is advisable to define the general principles to be adopted when measuring work as well as the actual method of measurement.

If the Works relate to cables, overhead powerlines, above ground or underground pipelines, instrumentation, insulation and/or the Contractor undertakes to provide building or civil engineering work, some or all of the following provisions may be considered appropriate.

The sub-clause numbers used below relate to the sub-clause numbers used in the general conditions.

1. SUB-CLAUSE 5.2 - SITE DATA

Delete the existing sub-clause and replace with the following.

5.2 The Tender shall be deemed to have been based on such data on climatic, hydrological, soil and general conditions of the Site and for the operation of the Works as the Purchaser or the Engineer has made available in writing to the Contractor for the purposes of the Tender. The Contractor shall be responsible for his own interpretation of such data. Unless otherwise provided in the Specification, the Contractor shall not be deemed to have allowed in the Tender for the removal or diversion of mains, sewers, drains, conduits and similar services. If the Engineer gives any instructions to the Contractor for the removal, alteration or diversion thereof, the Contractor shall be entitled to treat such instruction as a variation under clause 27 (Variations) and any work done in compliance with such instruction shall be valued and paid for accordingly.

Copyright © 2000 Institution of Electrical Engineers

2. SUB-CLAUSE 5.7 - UNEXPECTED SITE CONDITIONS

Delete the existing sub-clause and replace with the following.

5.7 If during the progress of the Works any unexpected physical condition (other than weather or conditions due to weather) or artificial obstruction is encountered by the Contractor and such condition or obstruction could not reasonably have been ascertained from an inspection of the Site by the Contractor before he prepared the Tender or from information made available to the Contractor for the purposes of the Tender, the Contractor shall forthwith inform the Engineer of the condition or obstruction encountered and obtain the approval of the Engineer of the steps he proposes to take to deal with the same. If the Contractor in taking such steps incurs extra Cost, such Cost shall be added to the Contract Price.

3. SUB-CLAUSE 6.1 - STATUTORY AND OTHER REGULATIONS

Delete the existing sub-clause and replace with the following.

6.1 If the Cost to the Contractor of performing his obligations under the Contract shall be increased or reduced by reason of the making after the date of the Tender of any law or of any order, regulation or by-law having the force of law that shall affect the Contractor in the performance of his obligations under the Contract, or by any conditions referred to in sub-clause 11.2 (Wayleaves, consents, etc.) or by any restriction or obligation referred to in sub-clause 11.9 (Third party interests), the existence and nature of which was not fully disclosed to the Contractor before the date of the Tender, the amount of such increase or reduction shall be added to or deducted from the Contract Price, as the case may be.

4. SUB-CLAUSE 11.2 - WAYLEAVES, CONSENTS, ETC.

Delete the existing sub-clause and replace with the following.

11.2 The Purchaser shall, within the times stated in the Programme or, if not so stated, before the time specified for delivery of any Plant to the Site, obtain all consents, wayleaves and approvals in connection with the regulations and by-laws of any local or other authority which shall be applicable to the Works on the Site. The Purchaser shall inform the Contractor of any conditions attached to such consents, wayleaves and approvals in so far as the same may affect the execution of the Works. Unless otherwise provided in the Specification, the Purchaser shall give all notices required to be given by any such wayleave or consent or by any law or any order, regulation or bylaw of any national, local or other duly constituted authority in relation to the execution and completion of the Works and remedying of defects therein.

5. SUB-CLAUSE 11.9 - THIRD PARTY INTERESTS

Add the following new sub-clause.

11.9 In the performance of his obligations under the Contract, the Contractor shall observe all restrictions and perform all obligations in favour of or owed to the owners for the time being of any interest in land in or adjoining which the Works are to be executed so far as such restrictions and obligations shall have been brought to the Contractor's notice. The Contractor shall not be liable to the Purchaser for any damage caused in the execution of the Works to the property of third parties which could not have been avoided by the exercise of reasonable care.

6. SUB-CLAUSE 16.3 - ERRORS IN DRAWINGS, ETC. SUPPLIED BY THIRD PARTIES

Add the following new sub-clause.

16.3 Where the Specification requires the Contractor to obtain drawings, information, records or particulars of the Site from third parties, the Contractor shall be responsible for the adequacy and interpretation of the information obtained but the Contractor shall not be responsible for any alterations or remedial work necessitated by reason of discrepancies, errors or omissions in drawings, information, records or particulars provided by any such third party, the accuracy of which a responsible contractor would accept in good faith.

7. STATUTORY AND OTHER REQUIREMENTS

Add the following new clause.

The Contractor shall in the performance of the Contract conform in all respects with the provisions of any law or any order, regulation or by-law of any national, local or other duly constituted authority that may in any way affect the execution or completion of the Works and the remedying of any defects therein and with the rules and regulations of all public bodies, companies and persons whose property rights are affected or in any way may be affected by the Works (herein called "the Regulations") and in particular the Contractor shall observe such of the restrictions and perform such of the obligations as may be imposed on the Purchaser in relation to the Works under the Regulations provided always that:-

(a) the Purchaser shall:-

either

(i) give details in the Specification of any obligations to be performed or restrictions to be observed and which are imposed upon the Purchaser under the Regulations,

or

(ii) provide assistance at the time of tendering, including if so requested by the Contractor the provision of copy documents at the Purchaser's expense, to enable the Contractor to ascertain the nature and extent of any such obligations or restrictions so imposed upon the Purchaser.

(b) if any provision of the Contract or any instructions of the Engineer would cause the Contractor to be in breach of the Regulations, the Engineer shall issue such instructions as may be necessary to ensure conformity with the Regulations and if in complying with any such instructions the Contractor shall incur extra Cost, the instruction shall be treated as a variation under clause 27 (Variations) and valued accordingly.

(c) the Purchaser shall be responsible for the payment of all fees required to be paid by the Regulations in so far as such fees are payable solely by reason of, and in connection with, the construction and completion of the Works.

(d) where the Contractor is not required by the Contract to undertake any permanent reinstatement or making good of the Site or any part thereof or of the means of access thereto, any obligation of the Contractor to maintain any interim restoration thereof required by the Contract or to comply with his responsibilities under sub-clause 18.1 (Fencing, guarding, lighting and watching) in respect of the particular part of the Site or of the access thereto, shall cease when the same is handed over by the Purchaser to others for permanent reinstatement, when permanent reinstatement commences, or three months after the Contractor has given notice to the Engineer of the completion of interim restoration, whichever is the earliest.

8. PAYMENT BY MEASUREMENT

Add the following new clause.

Where any part of the Works is to be paid for according to the quantity of Plant supplied or work done the following shall apply:-

(a) Such part of the Works shall be measured from time to time by the Engineer or by the Contractor. No such measurement shall be made by either of them without the other being afforded a reasonable opportunity of attending and agreeing the measurements. The Contractor shall at the request of the Engineer open up any part of the Works which may have been covered up without his having been afforded reasonable opportunity of measuring or agreeing the measurements thereof and the Contractor shall open up and restore the same at his own expense. All measurements shall be made in accordance with the provisions of the Specification respecting methods of measurement and if no method of measurement is therein specified the Engineer shall, after consultation with the Contractor, carry out such measurements in accordance with such method or methods as the Engineer shall determine are fair and reasonable in all the circumstances.

(b) The sum payable in respect of such part of the Works shall be ascertained according to the price or rate appropriate thereto and specified in the Contract. If no appropriate price or rate has been specified, the Engineer shall determine the price or rate applicable thereto which shall be a fair and reasonable price or rate having regard to any prices or rates that may be specified in the Contract for similar Plant or work.

9. REINSTATEMENT

Add the following new clause.

The Contractor shall not be responsible for the cost of new materials for the replacement of any surface materials (such as but not limited to flagstones, bricks, kerbs or setts) or any pipes, conduits or underground structures which, before lifting or when uncovered or exposed, are agreed by the Engineer to be in such condition as to be incapable of removal without damage or incapable of being replaced satisfactorily.

ADDITIONAL SPECIAL CONDITIONS FOR USE WHERE THE CONTRACT IS TO PROVIDE SECTIONAL COMPLETION AND DAMAGES FOR DELAY IN COMPLETION OF SECTIONS

Foreword

If it is desired to impose a separate Time for Completion for each Section of the Works and to apply liquidated damages to the failure to complete a Section, the following suggested clauses may be used.

The sub-clause numbers used below relate to the sub-clause numbers used in the general conditions.

1. SUB CLAUSE 33.1 - EXTENSION OF TIME FOR COMPLETION

Delete the existing sub-clause and replace with the following.

33.1 If, by reason of any variation ordered pursuant to clause 27 (Variations) or of any act or omission on the part of the Purchaser or the Engineer or of any industrial dispute or by reason of circumstances beyond the reasonable control of the Contractor arising after the acceptance of the Tender, the Contractor shall have been delayed in the completion of the Works or of any Section thereof, whether such delay occurs before or after the Time for Completion applicable thereto, then provided that the Contractor shall as soon as reasonably practicable have given to the Purchaser or the Engineer notice of his claim for an extension of time with full supporting details, the Engineer shall on receipt of such notice grant the Contractor from time to time in writing either prospectively or retrospectively such extension of the relevant Time for Completion of the Section or of the Works as may be reasonable.

2. SUB-CLAUSE 33.2 - DELAYS BY SUB-CONTRACTORS

Delete the existing sub-clause and replace with the following.

33.2 Any delay on the part of a Sub-Contractor which prevents the Contractor from completing the Works or a Section thereof within the Time for Completion applicable thereto shall entitle the Contractor to an extension thereof provided such delay is due to a cause for which the Contractor himself would have been entitled to an extension of time under sub-clause 33.1 (Extension of time for completion).

3. SUB-CLAUSE 34.1 - DELAY IN COMPLETION

Delete the existing sub-clause and replace with the following.

34.1 If the Contractor fails to complete the Works or any Section thereof in accordance with the Contract, save as regards his obligations under clauses 35 (Performance tests) and 36 (Defects liability), within the Time for Completion applicable thereto, or if no time be fixed, within a reasonable time, there shall be deducted from the Contract Price or paid to the Purchaser by the Contractor the percentage stated in the Appendix of the Contract Value of such parts of the Works or, as the case may be, of such parts of the Section as cannot in consequence of the said failure be put to the use intended for each week between the relevant Time for Completion and the actual date of completion. The amount so deducted or paid shall not exceed the maximum percentage stated in the Appendix of the Contract Value of such parts of the Works or of such parts of the Section, and such deduction or payment shall, subject to sub-clause 34.2 (Prolonged delay), be in full satisfaction of the Contractor's liability for the said failure.

ADDITIONAL SPECIAL CONDITIONS FOR USE IN CONTRACTS WHICH ARE SUBJECT TO THE HOUSING GRANTS, CONSTRUCTION AND REGENERATION ACT 1996 ("the Act")

The sub-clause numbers used below relate to the sub-clause numbers used in the general conditions.

Foreword

Where the Contract is a "Construction Contract" within the meaning of the Act, either party has the right to refer any dispute arising under the Contract to adjudication (as an addition, or alternative, to arbitration under clause 52 (Disputes and arbitration) or litigation). The following Special Conditions apply the adjudication rules contained in Part I of The Scheme for Construction Contracts or, as appropriate, of The Scheme for Construction Contracts (Scotland) or of The Scheme for Construction Contracts in Northern Ireland - ("the Scheme")- to any such adjudication. The Special Conditions also make provision for the effect of adjudication on Engineer's decisions, instructions and orders and on the provisions for arbitration. The further Special Conditions relating to payment and extensions of time reflect the provisions of the Act in the context of an MF/1 Contract as an alternative to the provisions of Part II of the Scheme.

The provisions of the Act are extended to Northern Ireland by the Construction Contracts (Northern Ireland) Order 1997 (the Order). Where the Contract is to be performed in Northern Ireland, the Order and its equivalent articles should be substituted for the Act and for the referenced sections of the Act.

(i) Engineer and engineer's representative

Delete sub-clause 2.6 (Disputing engineer's decisions, instructions and orders) and replace with the following sub-clause 2.6.A and renumber the reference to sub-clause 2.6 in sub-clause 52.1 (Notice of arbitration) accordingly.

Disputing engineer's decisions instructions and orders

2.6.A If the Contractor by notice to the Engineer within 21 days after receiving any decision, instruction or order of the Engineer in writing or written confirmation thereof under sub-clause 2.5 (Confirmation in writing), disputes or questions the same, giving his reasons for so doing, the Engineer shall within a further period of 21 days by notice to the Contractor and the Purchaser with reasons, confirm, reverse or vary such decision, instruction or order.

If either the Contractor or the Purchaser disagrees with such decision, instruction or order as confirmed, reversed or varied, he shall be at liberty to refer the matter to arbitration within a further period of 21 days. In the absence of such a reference to arbitration within the said period of 21 days such decision, instruction or order of the Engineer shall be final and binding on the parties, subject to the right of either party to refer the dispute relating to such decision, instruction or order to adjudication at any time under the provisions of the Housing Grants, Construction and Regeneration Act 1996.

In any case where such a dispute under this sub-clause is referred to adjudication, either party shall be at liberty to refer the adjudicator's decision to arbitration under clause 52 (Disputes and arbitration) within 21 days after the date of the adjudicator's decision and in the absence of such a reference to arbitration within the said period of 21 days the adjudicator's decision shall be final and binding on the parties.

Copyright © 2000 Institution of Electrical Engineers

(ii) Adjudication

The following new clause 52A should be added:-

52A.1 By virtue of Section 108(5) of the Housing Grants, Construction and Regeneration Act 1996 the adjudication rules contained in Part I of The Scheme for Construction Contracts /(Scotland) /in Northern Ireland (delete as appropriate) shall apply to all adjudications of disputes under the Contract.

52A.2   (a) The adjudicator shall be .... of ....

(b) If no adjudicator is nominated in (a) above, or if the nominated adjudicator is unwilling or unable to act, an adjudicator shall be nominated by (delete as appropriate):-

The Institution of Electrical Engineers.

The Institution of Mechanical Engineers.

(c) Where neither paragraph (a) nor (b) applies, or where the person named in (a) has already indicated that he is unwilling or unable to act and (b) does not apply, the party referring the dispute to adjudication shall request an adjudicator nominating body to select a person to act as adjudicator.

(iii) Certificates and payment

1.   **SUB-CLAUSE 39.1 - APPLICATION FOR PAYMENT**

Add the following:-

Unless otherwise provided, applications for interim certificates of payment may be made by the Contractor to the Engineer monthly.

2.   **SUB-CLAUSE 40 .1 - PAYMENT**

Add the following new sub-clause 40A.1:-

40A.1 The Purchaser may not make payment of any lesser sum than that certified in an interim certificate of payment or withhold payment of any sum that would otherwise be due under a certificate of payment unless he has within 5 days after the issue of the certificate of payment notified the Contractor of the amount he proposes to pay and the basis on which that amount was calculated and the provision of the Conditions that entitles the Purchaser to withhold the amount concerned.

(iv) Extension of time for completion

SUB-CLAUSE 33.1 - EXTENSION OF TIME FOR COMPLETION

Add the following:-

The Time for Completion shall be extended by any period during which the Contractor has, pursuant to the provisions of Section 112 of the Housing Grants, Construction and Regeneration Act 1996, suspended performance of the Contract.

(v) Form of Sub-Contract

1.   **CLAUSE 14 - PAYMENT**

Add the following new sub-clause 14.2:-

14.2   Notwithstanding any provision of the Eighth Schedule the Contractor shall not be entitled to make any payment to the Sub-Contractor conditional upon the Contractor having received payment under the Main Contract from a third party (including for the avoidance of doubt the Purchaser) unless that third party, or any other person payment by whom is under the Sub-Contract (directly or indirectly) a condition of payment by that third party, is insolvent. For the purpose of this sub-clause "insolvent" has the meaning assigned by Section 113 of the Housing Grants, Construction and Regeneration Act 1996.

2. CLAUSE 20 - ADJUDICATION

Add the following new clause 20:-

20.1 Unless otherwise specified in the Eleventh Schedule any adjudication of disputes between the Contractor and the Sub-Contractor shall be conducted by the adjudicator named in the Main Contract or, as appropriate, by an adjudicator nominated by the body named in the Main Contract and in accordance with the adjudication rules applicable under the Main Contract.

3. SUB-CONTRACT SCHEDULES

Add the following new schedule:-

ELEVENTH SCHEDULE

(If the Main Contract provisions do not apply.)

(i) Particulars of Adjudicator ....……................. , or, if no adjudicator is nominated, or if the nominated adjudicator is unable or unwilling to act, an adjudicator shall be nominated by:-

[Here insert Particulars of Person to Nominate Adjudicator]

(ii) [Here insert Particulars of Adjudication Rules or Adjudication Procedures]

ADDITIONAL SPECIAL CONDITIONS FOR USE IN CONTRACTS WHICH ARE SUBJECT TO THE CONTRACTS (RIGHTS OF THIRD PARTIES) ACT 1999

Foreword

If it is desired that third parties should not unintentionally have a benefit conferred upon them and become entitled to enforce terms of the Contract, the following suggested additional sub-clause to the general conditions may be used.

The sub-clause numbers used below relate to the sub-clause numbers used in the general conditions.

Assignment and sub-contracting

3.3 Unless otherwise provided, the Purchaser and the Contractor do not intend any term of the Contract to be enforceable by a party other than themselves.

(Also, amend clause 3 heading so that it reads **Assignment, sub-contracting and third party benefits**.)

The following new suggested sub-clause should be added to the Sub-Contract.

2.3 The Contractor and the Sub-Contractor do not intend any term of the Sub-Contract to be enforceable by a party other than themselves.

FORM OF SUB-CONTRACT

THIS AGREEMENT is made the .... day of ....

BETWEEN: .... PLC/Limited (Contractor)

whose registered office is at ....

....

(hereinafter called 'the Contractor') of the one part and

.... PLC/Limited (Sub-Contractor)

whose registered office is at ....

....

(hereinafter called 'the Sub-Contractor') of the other part

WHEREAS:

A. The Contractor has entered into a contract (hereinafter called 'the Main Contract') particulars of which are set out in the First Schedule hereto.

B. The Sub-Contractor has agreed to supply the plant and execute the works particulars of which are given in the Third Schedule hereto (which forms part of the Works to be executed by the Contractor under the Main Contract) upon the terms and conditions herein contained.

C. The Contractor has under the provisions of the Main Contract obtained such consents as are necessary from the Engineer appointed under the Main Contract to enable him to enter into this Agreement.

NOW IT IS HEREBY AGREED as follows:

Definitions

1.1 Except where the context otherwise requires the following expressions shall have the following meanings:

'the Main Contract' means the contract, particulars of which are given in the First Schedule hereto.

5 'the Sub-Contract' means this Agreement and the documents referred to in the Second Schedule hereto.

'the Sub-Contract Works' means all Sub-Contract Plant to be supplied and all work to be done by the Sub-Contractor which are described in the Third Schedule hereto.

10 'Sub-Contract Plant' means all plant to be supplied by the Sub-Contractor hereunder.

'Sub-Contract Price' means the sum specified in the Fourth Schedule hereto as payable to the Sub-Contractor for the Sub-Contract Works.

'Sub-Contract Value' means such part of the Sub-Contract Price, adjusted to give effect to such additions or deductions as provided for in the Sub-Contract as is properly apportionable to the Sub-Contract Plant or work in question. In determining the Sub-Contract Value the state, condition and topographical location of the Sub-Contract Plant, the amount of work done and all other relevant circumstances shall be taken into account.

The expressions 'Purchaser', 'Engineer', 'Engineer's Representative', 'Site', 'Works', 'Plant', 'Conditions' and 'Special Conditions' shall have the meanings respectively assigned to them under the Main Contract.

General

2.1 The Sub-Contractor shall carry out the Sub-Contract Works in accordance with the Sub-Contract and to the reasonable satisfaction of the Contractor and of the Engineer.

2.2 The Sub-Contractor shall not assign the whole or any part of the benefit of the Sub-Contract nor shall he sub-let the whole or any part of the Sub-Contract Works without the previous written consent of the Contractor. Nothing in this sub-clause shall affect any right of the Sub-Contractor to assign either absolutely or by way of charge any sum which is or may become due to him hereunder.

Main contract

3.1 The Conditions shall be deemed to be incorporated in this Agreement and as between the Contractor and the Sub-Contractor shall apply to the Sub-Contract Works as they apply to the Works. The Contractor shall provide the Sub-Contractor with a copy of the Main Contract other than the details of the Contractor's prices.

3.2 Unless the context otherwise requires, the provisions of the Main Contract shall apply to the Sub-Contract as if the Contractor were the Purchaser therein stated and the Sub-Contractor were the Contractor thereunder.

3.3 Where under the Main Contract and the Conditions any liability of the Contractor is to be determined or limited by reference to the Contract Price or Contract Value as defined therein, the liability of the Sub-Contractor hereunder shall be determined as if the expressions Sub-Contract Price and Sub-Contract Value were respectively substituted therefor.

3.4 Subject to sub-clauses 3.3 and 3.5 the Sub-Contractor shall indemnify the Contractor against every liability which the Contractor may incur to any other person whatsoever and against all claims, demands, proceedings, damages, costs and expenses made against or incurred by the Contractor by reason of any breach by the Sub-Contractor of the Sub-Contract, but in no case whatsoever (except claims resulting from death or injury to any person caused by the negligence of the Sub-Contractor for which no limit applies) shall the Sub-Contractor's liability hereunder exceed the Sub-Contract Price.

3.5 The Contractor shall indemnify and hold harmless the Sub-Contractor from and against any and all claims made against the Contractor in respect of which the Contractor is entitled to an indemnity from the Sub-Contractor under sub-clause 3.4 to the extent that such claims exceed the Sub-Contract Price (except for claims to which no limit applies under sub-clause 3.4).

3.6 The Contractor agrees and declares that every limitation and exclusion of liability of the Contractor contained in the Main Contract shall as between the Contractor and the Sub-Contractor extend to protect the Sub-Contractor, his servants or agents (with the exception of liability for death or personal injury caused by wilful or negligent acts or omissions).

Facilities

4.1 The Contractor shall provide for the use of the Sub-Contractor free of charge such facilities on the Site as may be specified in the Sub-Contract.

Copyright © 2000 Institution of Electrical Engineers

Access

5.1 The Contractor shall provide access for the Sub-Contractor to such part or parts of the Site, and the means of access thereto whether temporary or otherwise, as may be necessary for the Sub-Contractor to execute the Sub-Contract Works but such access shall not be exclusive to the Sub-Contractor.

5.2 The Sub-Contractor shall at all times permit the Engineer, the Engineer's Representative and the Contractor, his servants and agents (including any other person engaged for the purposes of the Works) during working hours to have reasonable access to the Sub-Contractor's premises, to the Sub-Contract Works and to all parts of the Site where any work is being done or Sub-Contract Plant is for the time being stored.

Completion

6.1 The Sub-Contractor shall complete the Sub-Contract Works within the time for completion thereof specified in the Fifth Schedule hereto. If by reason of any circumstance which entitles the Contractor to an extension of the Time for Completion of the Works under the Main Contract, or by reason of the ordering of any variation to the Sub-Contract Works, or by reason of any breach by the Contractor the Sub-Contractor shall be delayed in the execution of the Sub-Contract Works, then in any such case provided the Sub-Contractor shall have given without delay written notice to the Contractor of the circumstances giving rise to delay, the time for completion hereunder shall be extended by such period as may in all the circumstances be justified. The Sub-Contractor shall in all cases take such action as may be reasonable for minimising or mitigating the consequences of any such delay.

Delay in completion

7.1 If the Sub-Contractor fails to complete the Sub-Contract Works or any part thereof [other than his obligations under clause 13 (Defects liability)] within the time for completion or any extension thereof granted under clause 6 (Completion), there shall be deducted from the Sub-Contract Price, or the Sub-Contractor shall pay to the Contractor as and for liquidated damages, the percentage (stated in the Sixth Schedule) of the Sub-Contract Value of that part of the Sub-Contract Works as cannot in consequence of the delay be put to the use intended for each week between such time for completion and the actual date of completion, but in no case shall the total amount to be deducted or so paid exceed the maximum percentage of the Sub-Contract Price stated in the Sixth Schedule hereto. Such deduction or payment shall be in full satisfaction of and to the exclusion of any other remedy of the Contractor against the Sub-Contractor in respect of the Sub-Contractor's failure to complete within the time for completion of the Sub-Contract Works.

Progress of works

8.1 The Sub-Contractor shall comply with all instructions and decisions of the Engineer and of the Engineer's Representative which are notified and confirmed in writing to him by the Contractor.

8.2 The Contractor shall have the same powers to give instructions and decisions to the Sub-Contractor in relation to the Sub-Contract Works as the Engineer has in relation to the Works under the Main Contract and the Sub-Contractor shall comply with all such instructions and decisions. The powers of the Contractor under this sub-clause may be exercised by the Contractor whether or not the Engineer has given instructions and decisions to the Contractor in relation thereto under the Main Contract.

8.3 The extra cost incurred by the Sub-Contractor in complying with any instructions or decisions under sub-clauses 8.1 and 8.2 shall be added to the Sub-Contract Price together with a reasonable allowance for profit provided that such costs were not caused or contributed to by any act or omission by, or any breach of contract by, the Sub-Contractor.

Variations

9.1 The Sub-Contractor shall not make any variations to the Sub-Contract Works either by way of addition or omission except variations by the Engineer under the Main Contract and confirmed in writing to the Sub-Contractor by the Contractor or ordered in writing by the Contractor. The Sub-Contractor shall not act upon an unconfirmed order for the variation of the Sub-Contract Works which is received directly by him from the Purchaser or the Engineer. If the Sub-Contractor receives any such order direct, he shall forthwith inform the Contractor and supply him with a copy of such order if given in writing. The Sub-Contractor shall only act upon such order as directed in writing by the Contractor who shall give the Sub-Contractor directions thereon with all reasonable speed.

Valuation of variations

10.1 Variations to the Sub-Contract Works ordered by the Engineer and confirmed in writing by the Contractor or ordered by the Contractor shall be valued and determined in accordance with the principles contained in the Main Contract where appropriate by reference to the rates and prices specified in the Sub-Contract for similar or analogous work. If there are no such rates and prices or if they are not applicable then the Sub-Contractor shall be entitled to be paid such sum as is fair and reasonable in all the circumstances. If for the purposes of the Main Contract a variation of the Sub-Contract Works is also a variation of the Works which the Engineer requires to be measured for the purposes of the valuation thereof the Contractor shall permit the Sub-Contractor to attend any measurement made by the Engineer or on his behalf.

Returns and notices, etc.

11.1 Where the Contractor is required by the terms of the Main Contract to give any return, account or notice to the Engineer or to the Purchaser, the Sub-Contractor shall in relation to the Sub-Contract Works provide the Contractor with a similar return, account or notice in sufficient detail as will enable the Contractor to comply with the terms of the Main Contract.

Insurances

12.1 The Contractor shall in accordance with the provisions of sub-clauses 47.1 (Insurance of works) and 47.2 (Extension of works insurance) of the general conditions maintain in force the policy of insurance in respect of the Works details of which are given in Part 1 of the Seventh Schedule hereto and the Sub-Contractor shall be included as a co-insured thereunder. In the event that the Sub-Contract Works are destroyed or damaged in such circumstances that a claim is established in respect thereof under the said policy of insurance then the Sub-Contractor shall be paid the amount of such claim or the amount of his loss, whichever is the less, and shall apply such sum in replacing or repairing that which was destroyed or damaged. The Sub-Contractor shall observe and comply with the conditions contained in the said policy of insurance.

12.2 The Sub-Contractor shall effect insurance against such risks as are specified in Part II of the Seventh Schedule hereto and in such sums and for the benefit of such persons as is specified therein and shall maintain such insurance until he has finally performed his obligations under the Sub-Contract.

Defects liability

13.1 The Sub-Contractor shall until the expiry of the Defects Liability Period under the Main Contract be responsible for making good any defect in or damage to the Sub-Contract Works to the like extent as the Contractor is responsible to make good defects or damage under the Main Contract. If any damage made good by the Sub-Contractor under the provisions of this clause was caused by the act, neglect or default of the Contractor, his servants or agents the Sub-Contractor shall be entitled to be paid by the Contractor the reasonable costs of making good such damage together with a reasonable allowance for profit.

Payment

14.1 The Sub-Contractor shall be entitled to payment for the Sub-Contract Works in accordance with the provisions set out in the Eighth Schedule hereto.

Copyright © 2000 Institution of Electrical Engineers

Determination of the main contract

15.1 If the Main Contract is determined for any reason whatsoever (other than the default of the Sub-Contractor hereunder) then the Sub-Contractor shall be entitled to be paid in respect of the Sub-Contract Works sums calculated in accordance with the provisions of the Main Contract so far as applicable but at the rates and prices contained in the Sub-Contract as if the Contractor had terminated the Main Contract by reason of default of the Purchaser, less only such sums as may have been paid to the Sub-Contractor on account of the Sub-Contract Price. Such payment shall be without prejudice to the rights of either party in respect of any breach of the Sub-Contract committed by the other prior to such determination.

15.2 If the Main Contract is determined by the Purchaser in consequence of any breach of the Sub-Contract by the Sub-Contractor then the provisions of sub-clause 15.1 shall not apply but the rights of the Contractor and the Sub-Contractor hereunder shall be as if the Contractor had terminated the Sub-Contract under clause 16 (Sub-contractor's default).

Sub–contractor's default

16.1 The Contractor may, after having given 7 days' notice to the Sub-Contractor, expel the Sub-Contractor from the Site and/or terminate the Sub-Contract if the Sub-Contractor:

(a) assigns the Sub-Contract or sub-lets the whole of the Sub-Contract Works without the Contractor's previous consent, or

(b) has abandoned the Sub-Contract, or

(c) has suspended the progress of the Sub-Contract Works and has not resumed it within 28 days after receiving notice from the Contractor to proceed, or

(d) becomes bankrupt or insolvent, or has a receiving order made against him, or compounds with his creditors, or, being a corporation, commences to be wound up (not being a member's voluntary winding up for the purposes of amalgamation or reconstruction) or has an administration order made against him or carries on business under an administrator, a receiver, a manager or liquidator for the benefit of his creditors or any of them.

Any such expulsion and/or termination shall be without prejudice to any other rights or powers of the Contractor under the Sub-Contract.

16.2 The Contractor shall, as soon as possible after the date of expulsion or termination value the Sub-Contract Works at that date in accordance with the rates and prices set out in the Sub-Contract.

16.3 The Contractor shall not be liable to make any further payments to the Sub-Contractor except as follows:

The Contractor shall calculate:

(a) the cost reasonably incurred by the Contractor in completing the Sub-Contract Works, and

(b) the total amount which would have been payable to the Sub-Contractor if he had completed the Sub-Contract Works.

If the sum calculated under (a) when added to the total amount paid to the Sub-Contractor at the date of expulsion or termination exceeds the sum calculated under (b), the Sub-Contractor shall on demand pay the amount of such excess to the Contractor.

Copyright © 2000 Institution of Electrical Engineers

If there is no such excess the Sub-Contractor shall be paid the amount by which the value certified under sub-clause 16.2 exceeds the total amount paid to the Sub-Contractor at the date of expulsion or termination.

Applicable law

17.1 Unless otherwise stated in the Ninth Schedule, the Sub-Contract shall in all respects be governed by and interpreted in accordance with the laws of England and English law shall govern the procedure of any arbitration under clause 19 (Disputes and arbitration).

Jurisdiction

18.1 Subject to the provisions of this clause and clause 19 (Disputes and arbitration) below, any and all proceedings against the Contractor by the Sub-Contractor shall be brought in the courts of the country whose law governs the Sub-Contract to the jurisdiction of which the parties hereby agrees to submit. Any proceedings which may be brought by the Contractor against the Sub-Contractor may be brought either in the courts of the country whose law governs the Sub-Contract or in the courts of the country where the Sub-Contractor has his principal place of business.

Disputes and arbitration

19.1 If at any time any question, dispute or difference shall arise between the Contractor and the Sub-Contractor in relation to the Sub-Contract or in any way connected with the Sub-Contract Works, which cannot be settled amicably, either party shall as soon as is reasonably practicable give to the other notice of the existence of such question, dispute or difference specifying its nature and the point at issue, and the same shall be referred to the arbitration of a person to be agreed upon. Failing agreement upon such person within 30 days after the date of such notice, the arbitration shall be conducted by some person appointed on the application of either party by the President of the institution named in the Tenth Schedule (or by his deputy appointed by such President for the purpose). In any such arbitration, any decision of the Engineer which has become final and binding on the Contractor and the Purchaser under sub-clause 2.6 (Disputing engineer's decisions, instructions and orders) of the Main Contract shall also be and be treated as final and binding upon the Contractor and the Sub-Contractor.

19.2 The arbitrator shall have power:

(a) to open up, review and revise any certificate or valuation of the Engineer or any instruction or decision of the Engineer given to the Sub-Contractor by the Contractor under sub-clause 8.1 which has not become final and binding on the Contractor and the Sub-Contractor;

(b) to open up, review and revise any instruction or decision of the Contractor given to the Sub-Contractor under sub-clause 8.2;

(c) to order on a provisional basis (subject to the arbitrator's final award) any relief which he would have power to grant in a final award including but not limited to the making of a provisional order for the payment of money as between the parties or an order to make an interim payment on account of the costs of the arbitration.

19.3 If any question, dispute or difference between the Contractor and the Sub-Contractor which arises at any time before the expiry of three months after the issue of the final certificate of payment under the Main Contract or within three months after the termination of the Main Contract shall be substantially the same as a matter which is a question, dispute or difference between the Contractor and the Purchaser and/or the Engineer which has been submitted to arbitration under the Main Contract, either the Contractor or the Sub-Contractor shall be entitled to require the Sub-Contractor to be joined as a party to such arbitration. The Sub-Contractor hereby agrees to be so joined and to the reference of the dispute to the arbitrator appointed under the Main Contract in accordance with the arbitration rules applicable to the dispute under the Main Contract and further agrees that the arbitrator shall have power to order the consolidation of such arbitration proceedings and/or to order the holding of concurrent hearings.

Copyright © 2000 Institution of Electrical Engineers

19.4 Unless otherwise specified in the Tenth Schedule any arbitration under the provisions of this clause shall be conducted in accordance with the arbitration rules specified in the Main Contract and if none be so specified in accordance with the provisions of the Arbitration Act 1996.

19.5 Performance of the Sub-Contract shall continue during arbitration proceedings unless otherwise agreed by the parties or, in the case of an arbitration under sub-clause 19.3, unless the Contractor is otherwise instructed under the Main Contract by the Engineer.

AS WITNESS THE HANDS OF THE PARTIES

Note: If it is agreed that Scots law shall govern the Sub-Contract, similar amendments to those recommended for the Main Contract should also be made to the equivalent provision of the Sub-Contract.

FIRST SCHEDULE
Particulars of Main Contract

SECOND SCHEDULE
Particulars of Sub-Contract Documents

THIRD SCHEDULE
Particulars of Sub-Contract Works

FOURTH SCHEDULE
Sub-Contract Price

FIFTH SCHEDULE
Time for Completion of Sub-Contract Works

SIXTH SCHEDULE

Damages

[....................]% of Sub-Contract Value for Each Week of Delay to be Deducted or Paid

[.................]% of Sub-Contract Price (Maximum Which Deductions or Payments Shall not Exceed).

SEVENTH SCHEDULE

PART 1
Particulars of Works Insurances Under the Main Contract

PART II
Particulars of Insurances to be Effected by Sub-Contractor

EIGHTH SCHEDULE
Terms of Payment

NINTH SCHEDULE

[If it is desired that the Sub-Contract should be subject to a law other than English law or to the same law as the Main Contract, the appropriate provision should be included here. The following clause, which is appropriate for use in all cases where the Sub-Contract is to be governed by the same law as the Main Contract, may be used in substitution for clause 17.1 (Applicable law).

Applicable law

17.1   The Sub-Contract shall be governed and interpreted in conformity with the law governing the Main Contract.]

TENTH SCHEDULE

Particulars of Person to Appoint Arbitrator

Particulars of Arbitration Rules

Copyright © 2000 Institution of Electrical Engineers

FORM OF TENDER

DATE: ....

TO:

....

....PLC/Limited, (Purchaser)

Sirs,

[**SHORT DESCRIPTION OF THE WORKS**]

I/We, the undersigned, hereby tender and offer to design, manufacture, deliver, erect, test and complete the Works more particularly described and referred to in the general conditions and Special Conditions, Specification, schedules and drawings (if any) hereto annexed including addenda nos. [................................]
5 issued for this Tender and which under the terms thereof is to be designed, manufactured, delivered, erected, tested and completed by the Contractor and to perform and observe the provisions and agreements on the part of the Contractor contained in or reasonably to be inferred from the Conditions, Specification, schedules and drawings for the sum, exclusive of Value Added Tax,
10 of £ [............................], the details of which are given in the schedule of prices submitted herewith.

I/We further declare that I/we have visited and inspected the Site and have read and understood the tender documents.

I/We hereby undertake, in the event of your acceptance of this Tender and if
15 required, to execute the Agreement within 45 days from receipt of the Letter of Acceptance and if required to furnish a satisfactory Performance Bond in such amount as you may require not exceeding [10%] of the Contract Price and to obtain such insurance as is stipulated in the general conditions of Contract.

20 I/We undertake to do any extra work not covered by the above schedule of prices which may be ordered by the Engineer and hereby agree that the value of such extra work shall be determined as provided for in the general or Special Conditions of Contract.

I/We understand that you reserve to yourself the right to accept or refuse this
25 Tender whether it be lower, the same or higher than any other tender, or for any other reason.

I/We agree that this Tender shall remain open for acceptance by you and will not be withdrawn by us for a period of [..........................] days from the closing date for submission of tenders.

30 I/We hereby agree that the following schedules are attached and form part of this Tender:

Copyright © 2000 Institution of Electrical Engineers

Schedule of prices

[........................]

(Add further schedules as may be required, eg. Proposed Programme, Recommended Spare Parts, Plant Delivery Schedule, Performance Guarantees, Qualifications to Tender Documents.)

All correspondence relative to this Tender is to be addressed to the undersigned Tenderer at the following address:

....
....
....
....

The undersigned is empowered to sign this Tender on our behalf.

For and on behalf of
[Tenderer]

FORM OF AGREEMENT

This Agreement made the …………………… day of …………………… …………

BETWEEN

(1) …………………………………………………………… PLC/Limited (Purchaser)

of ……………………………………………………………………………………………,

(hereinafter called "the Purchaser") of the one part; and

(2) ……………………………………………………………PLC/Limited (Contractor)

of

……………………………………………………………………………………………,

(hereinafter called "the Contractor") of the other part.

WHEREAS the Purchaser wishes to have certain Works executed by the Contractor, viz.

[………………………………………………………………………………………]

and has appointed …………………………………………………… (Engineer)

of …………………………………………………………………………………… the Engineer for the purposes thereof (hereinafter called "the Engineer") and has accepted a Tender by the Contractor for the design, manufacture, delivery to Site, erection, testing and completion thereof and the remedying of defects therein in accordance with the general and Special Conditions hereinafter referred to under the direction of the Engineer in the sum of £ […………………] (hereinafter called "the Contract Price").

NOW THIS AGREEMENT WITNESSETH as follows:

1. In this Agreement words and expressions shall have the same meanings as are respectively assigned to them in the general conditions hereinafter referred to.

2. The following documents and their annexes which have been initialled by the parties and bound in herewith shall be deemed to form and be read and construed as part of this Agreement, viz:

    (a)   The general conditions.

    (b)   The Special Conditions (if any).

    (c)   The Specification and the drawings (if any) listed herein or annexed hereto.

    (d)   The Schedules (here set out [schedules of prices, Tests on Completion, Performance Tests, terms of payment, etc.]).

    (e)   The Letter of Acceptance.

    (f)   The said Tender.

3. In consideration of the payments to be made by the Purchaser to the Contractor as hereinafter mentioned the Contractor agrees to design, manufacture, deliver to Site, erect, test and complete the Works and to remedy defects therein in conformity in all respects with the provisions of the Contract.

Copyright © 2000 Institution of Electrical Engineers

4.   The Purchaser shall pay the Contractor in consideration of the execution and completion of the Works and the remedying of defects therein the Contract Price or such other sum as may become payable under the provisions of the Contract together with the Value Added Tax properly chargeable thereon at the times and in the manner prescribed by the Contract.

5.   If any question, dispute or difference shall arise between the Purchaser and the Contractor in relation to the Contract or in any way related to the Works, which cannot be settled amicably, it shall be referred to arbitration in accordance with clause 52 (Disputes and arbitration) of the general conditions.

Either (a)   IN WITNESS whereof the parties hereto have caused this Agreement to be entered into in the manner required by their respective constitutions and the laws of their respective countries.

Signed by [... (name) .....]
for and on behalf of the
Purchaser

[....] (Signature)

[....] (Title)

Signed by [... (name) .....]
for and on behalf of the
Contractor

[....] (Signature)

[....] (Title)

Or, (b)   if the Agreement is to be executed as a deed,

IN WITNESS whereof the parties have executed and delivered this Agreement as a deed on the date above written.

EXECUTED AND DELIVERED as a deed by

[.................................] PLC/LTD (Contractor)

in the presence of:

Director   [....] (Signature)

Director/Secretary [....] (Signature)

EXECUTED AND DELIVERED as a deed by

[.................................] PLC/LTD (Purchaser)

in the presence of:

Director   [....] (Signature)

Director/Secretary [....] (Signature)

Copyright © 2000 Institution of Electrical Engineers

FORM OF PERFORMANCE BOND

THIS PERFORMANCE BOND effective as of the date hereof is made as a deed BETWEEN the following parties whose names and registered offices/addresses are set out in the Schedule to this Bond ("the Schedule"):-

    (1)   the Contractor as Principal;

    (2)   the Guarantor as Guarantor; and

    (3)   the Purchaser as Beneficiary.

WHEREAS:

(1) By a contract ("the Contract") entered into or to be entered into between the Purchaser and the Contractor, particulars of which are set out in the Schedule, the Contractor has agreed with the Purchaser to design, manufacture, deliver, erect and test certain works ("the Works") and correct defects therein upon and subject to the terms of the Contract.

(2) The Guarantor has agreed with the Purchaser at the request of the Contractor to guarantee the performance of the obligations of the Contractor under the Contract upon the terms and conditions of this Performance Bond subject to the limitation set out in clause 2.

NOW THIS DEED WITNESSES as follows:-

1. The Guarantor guarantees to the Purchaser that in the event of Default by the Contractor (which shall include a termination of the Contract by the Purchaser under the provisions of clause 50.1(a) thereof) the Guarantor shall subject to the provisions of this Performance Bond satisfy and discharge the damages sustained by the Purchaser (or as appropriate discharge the sum deemed a debt due under the Contract from the Contractor to the Purchaser) as established and ascertained pursuant to and in accordance with the provisions of or by reference to the Contract and taking into account all sums due or to become due to the Contractor. The Guarantor shall not be entitled to perform the Contractor's obligations under the Contract.

2. The maximum aggregate liability of the Guarantor and the Contractor under this Performance Bond shall not exceed the sum set out in the Schedule ("the Bond Amount") but subject to such limitation and to clause 4 below the Guarantor shall have no greater liability to the Purchaser under this Performance Bond than the Contractor has to the Purchaser under the Contract.

3. The Guarantor shall not be discharged or released by any alteration of any of the terms, conditions and provisions of the Contract or in the extent or nature of the Works and no allowance of time by the Purchaser under or in respect of the Contract or the Works shall in any way release, reduce or affect the liability of the Guarantor under this Performance Bond.

4. Whether or not this Performance Bond shall be returned to the Guarantor the obligations of the Guarantor under this Performance Bond shall be released and discharged absolutely upon the Expiry Date (as conclusively defined in the Schedule to this Performance Bond).

5. This Performance Bond shall be governed by and construed in accordance with the laws of [....]* and any dispute under the Performance Bond shall be determined by the courts of the country

Copyright © 2000 Institution of Electrical Engineers

5   of such law. This Performance Bond incorporates and shall be subject to the Uniform Rules for Contract Bonds published by the International Chamber of Commerce (Publication No. 524) and words used in this Performance Bond shall have the meanings set out in the Rules but the provisions of paragraph (i) of Article 7(j) and the arbitration provisions of Article 8 shall not apply.

* Note: If not completed the law governing the Contract shall apply.

THE SCHEDULE

The Contractor: [.......................... (name) ....] whose registered office is at [.......................... (address) ....]

The Guarantor: [.......................... (name) ....] whose registered office is at [.......................... (address) ....]

The Purchaser: [.......................... (name) ....] whose registered office is at [.......................... (address) ....]

The Contract: A contract [dated the .... day of ....] [*to be entered into*] between the Contractor and the Purchaser (substantially in the form known as The Model Form of General Conditions of Contract MF/1) for the design, manufacture, delivery, erection and testing of [*short description of the Works*] for the contract sum of [*sum in words*] pounds sterling (or appropriate currency) (£ [..............]).

The Bond Amount: The sum of [........................ (sum in words) ....] pounds sterling (or appropriate currency) (£ [....]).

[Insert any provisions for reduction of the Bond Amount.]

Expiry Date: The date which is three calendar months after the date of issue of the last final certificate of payment under the Contract unless prior to the expiry of such period either the Contractor or the Purchaser shall have commenced proceedings arising out of the Contract in which event three calendar months after such proceedings have been finally concluded, or [Insert details of the date agreed between the parties as constituting the Expiry Date or of the method by which that date is to be determined.].

IN WITNESS whereof the Contractor and the Guarantor have executed and delivered this Performance Bond as a deed this .... day of ....

EXECUTED AND DELIVERED as a deed }
by the CONTRACTOR as Principal

Director [....] (Signature)

Director/Secretary [....] (Signature)

EXECUTED AND DELIVERED as a deed }
by the GUARANTOR as Guarantor

Director [....] (Signature)

Director/Secretary [....] (Signature)

Copyright © 2000 Institution of Electrical Engineers

FORM OF DEFECTS LIABILITY DEMAND GUARANTEE

THIS DEFECTS LIABILITY DEMAND GUARANTEE effective as of the date hereof is made as a deed BETWEEN the following parties whose names and registered offices /addresses are set out in the Schedule to this Guarantee ("the Schedule"):-

(1) the Contractor as Principal;

(2) the Guarantor as Guarantor; and

(3) the Purchaser as Beneficiary.

WHEREAS:

(1) By a contract ("the Contract") entered into between the Purchaser and the Contractor, particulars of which are set out in the Schedule, the Contractor agreed to design, manufacture, deliver, erect and test certain works ("the Works").

(2) Under the terms of the Contract the Contractor has agreed to assume certain obligations in relation to defects and damage caused by defects which may appear or occur in the Works during the Defects Liability Period defined in the Contract.

(3) The Purchaser has pursuant to the express terms of the Contract agreed that no part of the Contract Price payable under the Contract shall be withheld to secure the Contractor's obligations in relation to such defects and damage.

NOW THIS DEED WITNESSES as follows:-

1. The Guarantor hereby irrevocably undertakes to pay to the Purchaser any sum or sums not exceeding in total the amount of £ [...........................] (insert total value of retention which would otherwise be retained) upon receipt by the Guarantor of the Purchaser's written demand.

2. The Purchaser's written demand shall be accompanied by a written statement certifying either that the Contractor has failed to carry out his obligations to rectify certain specified defects or damage for which he is responsible under the Contract during the said Defects Liability Period and the nature of such defects or damage or by a certificate that the Purchaser has become entitled to terminate the Contract under clause 50.1(a) thereof and the circumstances which gave rise thereto.

3. The Guarantor's liability under this Guarantee shall not exceed the total amount of retention money released to the Contractor by the Purchaser evidenced by certificates of payment issued under the terms of the Contract and copied to the Guarantor.

4. The Purchaser's demand for payment and written statement must be under the Purchaser's signature(s) authenticated by the Purchaser's bankers.

5. Any written demand and statement must be received by the Guarantor at the address stated in the Schedule on or before [...........................] ("the Expiry Date") / the date the Guarantor receives a copy of the final certificate of payment under the Contract ("the Expiry Event") when this

Guarantee shall expire. On expiry of the Guarantee the Purchaser shall return the Guarantee to the Guarantor.

6. This Guarantee is personal to the Purchaser and is not assignable.

7. This Guarantee shall be governed by the laws of [................................]* and shall be subject to the Uniform Rules for Demand Guarantees published by the International Chamber of Commerce (Publication No. 458).

* Note: If not completed the law governing the Contract shall apply.

THE SCHEDULE

The Contractor: [........................ (name) ....] whose registered office is at
[........................ (address) ....]

The Guarantor: [........................ (name) ....] whose registered office is at
[........................ (address) ....]

The Purchaser: [........................ (name) ....] whose registered office is at
[........................ (address) ....]

The Contract: A contract [dated the .... day of ....] between the Contractor and the Purchaser (substantially in the form known as The Model Form of General Conditions of Contract MF/1) for the design, manufacture, delivery, erection and testing of [*short description of the Works*] for the contract sum of [*sum in words*] pounds sterling (or appropriate currency) (£ [.............]).

IN WITNESS whereof the Contractor and the Guarantor have executed and delivered this Guarantee as a deed this .... day of ....

EXECUTED AND DELIVERED as a deed }
by the CONTRACTOR as Principal

Director [....] (Signature)

Director/Secretary [....] (Signature)

EXECUTED AND DELIVERED as a deed }
by the GUARANTOR as Guarantor

Director [....] (Signature)

Director/Secretary [....] (Signature)

Copyright © 2000 Institution of Electrical Engineers

FORM OF NOTICE OF DELEGATION OF AUTHORITY

(Including notice of appointment of Engineer's Representative)

[....] (Engineer)

As Engineer appointed under

[....] (Title of Contract) -"the Contract"-

Hereby gives you notice under sub-clause 2.3 (Engineer's power to delegate) of the general conditions that the following duties of the Engineer under the Contract in relation to the Works/Section of the Works (*as appropriate*) noted below are delegated to

[....] who is appointed as Engineer's Representative.

Sub-Clause Number	Title	Duties Delegated to the Engineer's Representative
5.7	Unexpected site conditions	Approving Contractor's steps
14.1	Programme	Approval of Programme
(*Continue with identification of the duties and the extent of the delegation and the sub-clauses under which the duties arise as appropriate to the Contract and the particular delegation.*)		

This notice supersedes all previous delegation of duties to the Engineer's Representative named above *(or as appropriate)/ revokes and supersedes/ is in addition to the delegation dated [...............]/*

Signed [....] (Engineer)

Works/Section to which this notice relates [....

....]

Date [....]

Copyright © 2000 Institution of Electrical Engineers

VARIATION ORDER

DATE [....................]

V.O. No. [....................]

[SHORT DESCRIPTION OF WORKS]

TO: .... PLC/Limited (Contractor)

You are directed under clause 27 (Variations) of the general conditions to make the variation to the Works noted below:

(1) .... (2) ....

Signed (Purchaser (where required))   Signed (Engineer)

Dated ....   Dated ....

Nature of the changes: ....

....

Enclosures: ....

....

----

1. This variation results in the following adjustment of the Contract Price in accordance with clause 27 (Variations):

Contract Price   £ [..................................]

Net increase/decrease resulting
from previous variation orders   £ [..................................]

Net increase/decrease resulting
from this variation order   £ [..................................]

Total sum now due to Contractor   £ [..................................]

2. The Time for Completion shall be adjusted as follows:

Time for Completion prior
to this variation order   [..........................] (period of days, or date)

Net increase resulting from this variation order [....................] (days)

Revised Time for Completion [....................] (period of days, or date)

3. The adjustments made under paragraph 1 make no allowance for any variation in Cost to which the Contractor may be entitled under clause 6 (Changes in costs) of the general conditions.

4. The Contractor's obligations under the Contract otherwise than in relation to Time for Completion shall be modified as follows:

[....]
(provide details)

Copyright © 2000 Institution of Electrical Engineers

TAKING-OVER CERTIFICATE

TO: [**CONTRACTOR**]

[**PURCHASER**]

[**SHORT DESCRIPTION OF WORKS**]

1. In accordance with clause 29 (Taking-over) of the general conditions of Contract it is hereby certified that the Works specified in the Schedule below was completed (except in minor respects that do not affect its commercial use, noted below) and has passed the Tests on Completion and is accordingly deemed to have been taken over by the Purchaser with effect from

[....] (Taking-Over Date).

SCHEDULE

[Works to which this certificate applies]

2. Accordingly, the Defects Liability Period in respect of the Works specified in the Schedule commenced on [.....................] (Taking-Over Date) and subject to clause 36 (Defects liability) of the general conditions will expire on   [....].

3. The items specified in the list annexed hereto remain to be completed or corrected. The Contractor is instructed to complete or correct the same within 28 days after the date of this certificate.

4. Retention of £ [............................] due on issue of the Taking-Over Certificate shall be paid by the Purchaser to the Contractor within [28] days after the date hereof.

5. Having regard to claims for extensions of time under sub-clause 33.1 (Extension of time for completion) received prior to the date hereof it is certified that for the purposes of clause 34 (Delay) the Time for Completion expired on [.............................].

ENGINEER .... Date ....

Index to general conditions

	Clause
Access to site	11.1
accident, etc. to workmen, Insurance against	47.5
Accidents or injury to workmen	43.6
act fairly, Engineer to	2.7
Additional cost caused by suspension	25.2
additional cost or payment, Disallowance of	25.4
Adjustments and modifications	35.4
adjustments and modifications, Postponement of	35.5
Adjustments to certificates	39.5
agents, Sub-contractors, servants and	53.1
Agreement	7.1
Allowance for profit on claims	41.2
Alterations to programme	14.4
Applicable law	54.1
Application for final certificate of payment	39.9
Application for payment	39.1
Application of insurance monies	47.3
application, Form of	39.2
Approval of programme	14.3
Approved drawings	15.3
Arbitration rules	52.5
arbitration, Notice of	52.1
arbitration, Performance to continue during	52.2
Arbitrator's powers	52.3
Assignment	3.1
Assistance with laws and regulations	12.1
Bankruptcy and insolvency	50.1
bond or guarantee, Failure to provide	8.2
bond or guarantee, Provision of	8.1
Breach of purchaser's general obligations	11.8
Care of the works	43.1
certificate of payment, Application for final	39.9
certificate of payment, Effect of final	39.12
certificate of payment, Issue of final	39.11
certificate of payment, Value of final	39.10
certificate of payment, Withholding	39.7
Certificate of test or inspection	23.4
certificate, Effect of taking-over	29.3
certificate, Issue of payment	39.3
certificate, Taking-over	29.2
certificates of payment, Effect of	39.8
certificates of payment, Value included in	39.4
certificates, Adjustments to	39.5
certificates, Corrections to	39.6
certify or make payment, Remedies on failure to	40.3
Cessation of performance tests	35.3

	Clause
Claims in respect of injury to persons or damage to property	43.7
claims, Allowance for profit on	41.2
claims, Extraordinary traffic	21.3
claims, Notification of	41.1
claims, Purchaser's liability to pay	41.3
Clearance of site	18.3
completion of performance tests, Time for	35.6
completion, Consequences of failure to pass tests on	28.5
completion, Delay in	34.1
completion, Extension of time for	33.1
completion, Time for	32.1
conditions, Unexpected site	5.7
Conduct of proceedings	42.2
confidential, Details	9.1
Confirmation in writing	2.5
confirmation of variations, Notice and	27.5
consents, etc., Wayleaves	11.2
consents, Notices and	1.4
consequences of delay, Mitigation of	33.3
Consequences of disapproval of drawings	15.2
Consequences of failure to pass performance tests	35.8
Consequences of failure to pass tests on completion	28.5
consequential damage, Indirect or	44.2
contractor has responsibility for care of the works, Injury to persons or damage to property whilst	43.4
Contractor to inform himself fully	5.1
Contractor to search	36.8
contractor, Errors in drawings, etc. supplied by	16.1
contractor, Notices to	10.2
contractor, Purchaser's use of drawings, etc. supplied by	15.7
contractors, Opportunities for other	18.4
Contractor's default	49.1
Contractor's design	13.3
Contractor's equipment	38.1
Contractor's equipment on site	38.2
contractor's equipment, Loss or damage to	38.3
contractor's equipment, Maintenance of	38.4
contractor's equipment, Removal of	51.2
Contractor's general obligations	13.1
contractor's liability, Limitation of	44.3
Contractor's records of costs	27.4
Contractor's representatives and workmen	17.1
Contractor's use of drawings, etc. supplied by purchaser or engineer	15.8

	Clause
Corrections to certificates	39.6
cost caused by suspension, Additional	25.2
cost or payment, Disallowance of additional	25.4
costs, Contractor's records of	27.4
cover, Exclusions from insurance	47.7
damage to contractor's equipment, Loss or	38.3
damage to property, Claims in respect of injury to persons or	43.7
damage to property after responsibility for care of the works passes to purchaser, Injury to persons or	43.5
damage to property whilst contractor has responsibility for care of the works, Injury to persons or	43.4
damage to the works, Making good loss or	43.2
damage, Indirect or consequential	44.2
data, Site	5.2
data, Site	5.3
Date for test or inspection	23.2
date of termination, Valuation at	49.2
decisions, instructions and orders, Disputing engineer's	2.6
decisions, instructions and orders, Engineer's	2.4
default, Contractor's	49.1
default, Notice of termination due to purchaser's	51.1
default, Payment on termination due to purchaser's	51.3
defective work, Removal of	36.6
Defects after taking-over	36.1
Defects before taking-over	26.1
defects liability period, Tests during	31.2
defects liability, Effect of suspension on	25.6
defects liability, Extension of	36.4
defects, Delay in remedying	36.5
defects, Latent	36.10
defects, Limitation of liability for	36.9
defects, Making good	36.2
defects, Notice of	36.3
Definitions	1.1
Delay in completion	34.1
Delay in remedying defects	36.5
delay, Mitigation of consequences of	33.3
delay, Prolonged	34.2
Delayed payment	40.2
Delayed tests	28.3
Delays by sub-contractors	33.2
delegate, Engineer's power to	2.3
Delivery	24.1

	Clause
delivery or erection, Resumption of work,	25.5
design, Contractor's	13.3
Details confidential	9.1
Disallowance of additional cost or payment	25.4
disapproval of drawings, Consequences of	15.2
Disputing engineer's decisions, instructions and orders	2.6
documents, Precedence of	4.1
Drawings	15.1
drawings, Approved	15.3
drawings, Consequences of disapproval of	15.2
drawings, etc., Manufacturing	15.9
drawings, etc. supplied by contractor, Errors in	16.1
drawings, etc. supplied by contractor, Purchaser's use of	15.7
drawings, etc. supplied by purchaser or engineer, Contractor's use of	15.8
drawings, etc. supplied by purchaser or engineer, Errors in	16.2
drawings, Foundation, etc.	15.5
drawings, Inspection of	15.4
duties, Engineer's	2.1
duties, Import permits, licences and	11.3
Effect of certificates of payment	39.8
Effect of final certificate of payment	39.12
Effect of suspension on defects liability	25.6
Effect of taking-over certificate	29.3
Engineer to act fairly	2.7
engineer, Contractor's use of drawings, etc. supplied by purchaser or	15.8
engineer, Errors in drawings, etc. supplied by purchaser or	16.2
engineer, Notices to purchaser and	10.1
engineer, Replacement of	2.8
Engineer's decisions, instructions and orders	2.4
engineer's decisions, instructions and orders, Disputing	2.6
Engineer's duties	2.1
Engineer's entitlement to test	23.1
Engineer's power to delegate	2.3
Engineer's power to vary	27.2
Engineer's representative	2.2
entitlement to test, Engineer's	23.1
equipment on site, Contractor's	38.2
equipment, Contractor's	38.1
equipment, Loss or damage to contractor's	38.3
equipment, Maintenance of contractor's	38.4
equipment, Purchaser's lifting	11.5
equipment, Removal of contractor's	51.2
erection, Resumption of work, delivery or	25.5
Errors in drawings, etc. supplied by contractor	16.1

	Clause
Errors in drawings, etc. supplied by purchaser or engineer	16.2
Evaluation of results of performance tests	35.7
Exclusions from insurance cover	47.7
Exclusive remedies	44.4
execution, Manner of	13.2
Extension of defects liability	36.4
Extension of time for completion	33.1
Extension of works insurance	47.2
Extraordinary traffic	21.1
Extraordinary traffic claims	21.3
Failure on test or inspection	23.5
failure to certify or make payment, Remedies on	40.3
failure to insure, Remedy on	48.1
failure to pass performance tests, Consequences of	35.8
failure to pass tests on completion, Consequences of	28.5
Failure to provide bond or guarantee	8.2
fairly, Engineer to act	2.7
Fencing, guarding, lighting and watching	18.1
final certificate of payment, Application for	39.9
final certificate of payment, Effect of	39.12
final certificate of payment, Issue of	39.11
final certificate of payment, Value of	39.10
Force majeure	46.1
force majeure, Notice of	46.2
force majeure, Payment on termination for	46.4
force majeure, Termination for	46.3
Form of application	39.2
Form of programme	14.2
Foundation, etc. drawings	15.5
Foundations, etc.	11.4
Further tests	36.7
General insurance requirements	47.6
general obligations, Breach of purchaser's	11.8
general obligations, Contractor's	13.1
gross misconduct, No effect in the case of	39.13
guarantee, Failure to provide bond or	8.2
guarantee, Provision of bond or	8.1
guarding, lighting and watching, Fencing	18.1
Headings and marginal notes	1.5
Hours of work	19.1
Import permits, licences and duties	11.3
Indemnity against infringement	42.1
indemnity against infringement, Purchaser's	42.3
Indirect or consequential damage	44.2
Infringement preventing performance	42.4
infringement, Indemnity against	42.1
infringement, Purchaser's indemnity against	42.3
Injury to persons or damage to property after responsibility for care of the works passes to purchaser	43.5
Injury to persons or damage to property whilst contractor has responsibility for care of the works	43.4
injury to persons or damage to property, Claims in respect of	43.7
injury to workmen, Accidents or	43.6
insolvency, Bankruptcy and	50.1
Inspection of drawings	15.4
inspection, Certificate of test or	23.4
inspection, Date for test or	23.2
inspection, Failure on test or	23.5
inspection, Services for test or	23.3
instructions and orders, Disputing engineer's decisions,	2.6
instructions and orders, Engineer's decisions,	2.4
Instructions to suspend	25.1
instructions, Operating and maintenance	15.6
Insurance against accident, etc. to workmen	47.5
insurance cover, Exclusions from	47.7
insurance monies, Application of	47.3
Insurance of works	47.1
insurance requirements, General	47.6
insurance, Extension of works	47.2
insurance, Third party	47.4
insurances, Joint	48.2
insure, Remedy on failure to	48.1
Interference with tests	31.1
Interpretation	1.2
Issue of final certificate of payment	39.11
Issue of payment certificate	39.3
items, Prime cost	5.5
items, Prime cost	5.6
Joinder	52.4
Joint insurances	48.2
Labour, materials and transport	6.2
labour, Returns of	17.3
Latent defects	36.10
law, Applicable	54.1
laws and regulations, Assistance with	12.1
liability for defects, Limitation of	36.9
liability to pay claims, Purchaser's	41.3
liability, Limitation of contractor's	44.3
licences and duties, Import permits,	11.3
lifting equipment, Purchaser's	11.5
lighting and watching, Fencing, guarding	18.1
Limitation of contractor's liability	44.3

	Clause
Limitation of liability for defects	36.9
loads, Special	21.2
Loss or damage to contractor's equipment	38.3
loss or damage to the works, Making good	43.2
loss, Mitigation of	44.1
maintenance instructions, Operating and	15.6
Maintenance of contractor's equipment	38.4
majeure, Force	46.1
majeure, Notice of force	46.2
majeure, Payment on termination for force	46.4
majeure, Termination for force	46.3
Making good defects	36.2
Making good loss or damage to the works	43.2
Manner of execution	13.2
Manufacturing drawings, etc.	15.9
marginal notes, Headings and	1.5
Marking of plant	37.2
materials and transport, Labour,	6.2
Meaning of variation	27.1
misconduct, No effect in the case of gross	39.13
Mitigation of consequences of delay	33.3
Mitigation of loss	44.1
modifications, Adjustments and	35.4
modifications, Postponement of adjustments and	35.5
monies, Application of insurance	47.3
night or rest day working, No	19.2
No effect in the case of gross misconduct	39.13
No night or rest day working	19.2
notes, Headings and marginal	1.5
Notice and confirmation of variations	27.5
Notice of arbitration	52.1
Notice of defects	36.3
Notice of force majeure	46.2
Notice of termination due to purchaser's default	51.1
Notice of tests	28.1
Notices and consents	1.4
Notices to contractor	10.2
Notices to purchaser and engineer	10.1
notices, Service of	10.3
Notification of claims	41.1
Objection to representatives	17.2
obligations, Breach of purchaser's general	11.8
obligations, Contractor's general	13.1
Operating and maintenance instructions	15.6
Opportunities for other contractors	18.4
orders, Disputing engineer's decisions, instructions and	2.6
orders, Engineer's decisions, instructions and	2.4

	Clause
other regulations, Statutory and	6.1
Outstanding work	29.4
Ownership of plant	37.1
Payment	40.1
Payment after termination	49.3
payment certificate, Issue of	39.3
Payment for plant affected by suspension	25.3
Payment on termination due to purchaser's default	51.3
Payment on termination for force majeure	46.4
payment, Application for	39.1
payment, Application for final certificate of	39.9
payment, Delayed	40.2
payment, Disallowance of additional cost or	25.4
payment, Effect of certificates of	39.8
payment, Effect of final certificate of	39.12
payment, Issue of final certificate of	39.11
payment, Remedies on failure to certify or make	40.3
payment, Value included in certificates of	39.4
payment, Value of final certificate of	39.10
payment, Withholding certificate of	39.7
performance tests, Cessation of	35.3
performance tests, Consequences of failure to pass	35.8
performance tests, Evaluation of results of	35.7
performance tests, Procedures for	35.2
performance tests, Time for	35.1
performance tests, Time for completion of	35.6
Performance to continue during arbitration	52.2
performance, Infringement preventing	42.4
Period, Tests during defects liability	31.2
permits, licences and duties, Import	11.3
persons or damage to property, Claims in respect of injury to	43.7
plant affected by suspension, Payment for	25.3
plant, Marking of	37.2
plant, Ownership of	37.1
plural, Singular and	1.3
Postponement of adjustments and modifications	35.5
power to delegate, Engineer's	2.3
power to vary, Engineer's	27.2
powers, Arbitrator's	52.3
Power, etc. for tests on site	11.7
power, Utilities and	11.6
Precedence of documents	4.1
Prime cost items	5.5
Prime cost items	5.6
Procedures for performance tests	35.2
proceedings, Conduct of	42.2
profit on claims, Allowance for	41.2
Programme	14.1
programme, Alterations to	14.4

	Clause		Clause
programme, Approval of	14.3	responsibility for care of the works, Injury to persons or damage to property whilst contractor has	43.4
programme, Form of	14.2		
programme, Revision of	14.5		
Progress with variations	27.6	responsibility for care of the works passes to purchaser, Injury to persons or damage to property after	43.5
progress, Rate of	14.6		
Prolonged delay	34.2		
property after responsibility for care of the works passes to purchaser, Injury to persons or damage to	43.5	rest day working, No night or	19.2
		results of performance tests, Evaluation of	35.7
		Resumption of work, delivery or erection	25.5
		Returns of labour	17.3
property whilst contractor has responsibility for care of the works, Injury to persons or damage to	43.4	Revision of programme	14.5
		risks, Damage to works caused by purchaser's	43.3
		risks, Purchaser's	45.1
property, Claims in respect of injury to persons or damage to	43.7	rules, Arbitration	52.5
		Safety	20.1
provide bond or guarantee, Failure to	8.2	search, Contractor to	36.8
Provisional sums	5.4	sections, Taking-over by	29.1
Provision of bond or guarantee	8.1	servants and agents, Sub-contractors,	53.1
purchaser and engineer, Notices to	10.1	Service of notices	10.3
purchaser or engineer, Contractor's use of drawings, etc. supplied by	15.8	Services for test or inspection	23.3
		services, Site	18.2
purchaser or engineer, Errors in drawings, etc. supplied by	16.2	Setting out	22.1
		Singular and plural	1.3
purchaser, Injury to persons or damage to property after responsibility for care of the works passes to	43.5	site conditions, Unexpected	5.7
		Site data	5.2
		Site data	5.3
		Site services	18.2
purchaser's default, Notice of termination due to	51.1	site, Access to	11.1
		site, Clearance of	18.3
purchaser's default, Payment on termination due to	51.3	site, Contractor's equipment on	38.2
		site, Power, etc. for tests on	11.7
purchaser's general obligations, Breach of	11.8	Special loads	21.2
Purchaser's indemnity against infringement	42.3	Statutory and other regulations	6.1
Purchaser's liability to pay claims	41.3	Sub-contracting	3.2
Purchaser's lifting equipment	11.5	sub-contractors, Delays by	33.2
Purchaser's risks	45.1	Sub-contractors, servants and agents	53.1
purchaser's risks, Damage to works caused by	43.3	sums, Provisional	5.4
Purchaser's use of drawings, etc. supplied by contractor	15.7	suspend, Instructions to	25.1
		suspension on defects liability, Effect of	25.6
Rate of progress	14.6	suspension, Additional cost caused by	25.2
records of costs, Contractor's	27.4	suspension, Payment for plant affected by	25.3
regulations, Assistance with laws and	12.1	Taking-over by sections	29.1
regulations, Statutory and other	6.1	Taking-over certificate	29.2
Remedies on failure to certify or make payment	40.3	Taking-over certificate, Effect of	29.3
		Taking-over, Defects after	36.1
remedies, Exclusive	44.4	Taking-over, Defects before	26.1
Remedy on failure to insure	48.1	Taking-over, Use before	30.1
remedying defects, Delay in	36.5	termination due to purchaser's default, Notice of	51.1
Removal of contractor's equipment	51.2		
Removal of defective work	36.6	termination due to purchaser's default, Payment on	51.3
Repeat tests	28.4		
Replacement of engineer	2.8	Termination for force majeure	46.3
representative, Engineer's	2.2	termination for force majeure, Payment on	46.4
representatives and workmen, Contractor's	17.1	termination, Payment after	49.3
representatives, Objection to	17.2		
requirements, General insurance	47.6		

	Clause
termination, Valuation at date of	49.2
test, Engineer's entitlement to	23.1
test or inspection, Certificate of	23.4
test or inspection, Date for	23.2
test or inspection, Failure on	23.5
test or inspection, Services for	23.3
Tests during defects liability period	31.2
tests on completion, Consequences of failure to pass	28.5
tests on site, Power, etc. for	11.7
tests, Cessation of performance	35.3
tests, Consequences of failure to pass performance	35.8
tests, Delayed	28.3
tests, Evaluation of results of performance	35.7
tests, Further	36.7
tests, Interference with	31.1
tests, Notice of	28.1
tests, Procedures for performance	35.2
tests, Repeat	28.4
tests, Time for	28.2
tests, Time for completion of performance	35.6
tests, Time for performance	35.1
Third party insurance	47.4
Time for completion	32.1
Time for completion of performance tests	35.6
time for completion, Extension of	33.1
Time for performance tests	35.1
Time for tests	28.2
traffic claims, Extraordinary	21.3
traffic, Extraordinary	21.1
traffic, Waterborne	21.4
transport, Labour, materials and	6.2
Unexpected site conditions	5.7
Use before taking-over	30.1
use of drawings, etc. supplied by contractor, Purchaser's	15.7
use of drawings, etc. supplied by purchaser or engineer, Contractor's	15.8
Utilities and power	11.6
Valuation at date of termination	49.2
Valuation of variations	27.3
Value included in certificates of payment	39.4
Value of final certificate of payment	39.10
variation, Meaning of	27.1
variations, Notice and confirmation of	27.5
variations, Progress with	27.6
variations, Valuation of	27.3
vary, Engineer's power to	27.2
watching, Fencing, guarding, lighting and	18.1
Waterborne traffic	21.4

	Clause
Wayleaves, consents, etc.	11.2
Withholding certificate of payment	39.7
work, delivery or erection, Resumption of	25.5
work, Hours of	19.1
work, Outstanding	29.4
work, Removal of defective	36.6
working, No night or rest day	19.2
workmen, Accidents or injury to	43.6
workmen, Contractor's representatives and	17.1
workmen, Insurance against accident, etc. to	47.5
works caused by purchaser's risks, Damage to	43.3
works insurance, Extension of	47.2
works passes to purchaser, Injury to persons or damage to property after responsibility for care of the	43.5
works, Care of the	43.1
works, Injury to persons or damage to property whilst contractor has responsibility for care of the	43.4
works, Insurance of	47.1
works, Making good loss or damage to the	43.2
writing, Confirmation in	2.5